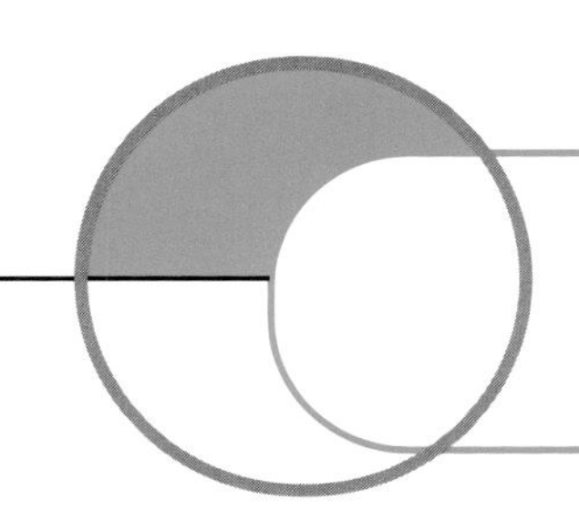

CONTENTS

PREFACE		vi
ABBREVIATIONS		vii
SECTION 1	**MEMBRANES AND LIPIDS**	1
SECTION 2	**CELL BIOLOGY**	37
SECTION 3	**GENES TO PROTEINS**	71
SECTION 4	**PROTEIN STRUCTURE AND FUNCTION**	109
INDEX		147

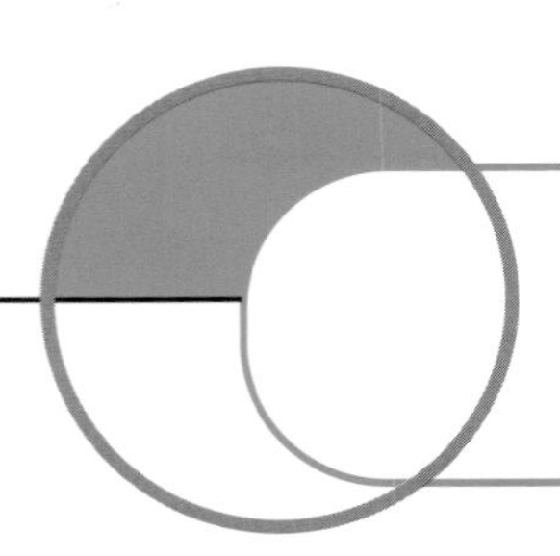

PREFACE

From the Series Editor, Elliott Smock

Are you ready to face your looming exams? If you have done loads of work, then congratulations; we hope this opportunity to practise SAQs, EMQs, MCQs and Problem-based Questions on every part of the core curriculum will help you consolidate what you've learnt and improve your exam technique. If you don't feel ready, don't panic – the One Stop Doc series has all the answers you need to catch up and pass.

There are only a limited number of questions an examiner can throw at a beleaguered student and this text can turn that to your advantage. By getting straight into the heart of the core questions that come up year after year and by giving you the model answers you need, this book will arm you with the knowledge to succeed in your exams. Broken down into logical sections, you can learn all the important facts you need to pass without having to wade through tons of different textbooks when you simply don't have the time. All questions presented here are 'core'; those of the highest importance have been highlighted to allow even sharper focus if time for revision is running out. In addition, to allow you to organize your revision efficiently, questions have been grouped by topic, with answers supported by detailed integrated explanations.

On behalf of all the One Stop Doc authors I wish you the very best of luck in your exams and hope these books serve you well!

From the Authors, Desikan Rangarajan and David Shaw

A good understanding of basic cell and molecular biology will be invaluable in the coming years of your future career. It may at the moment seem irrelevant, boring and sheer torture, but believe us when we say: learn the basic concepts now and your life will be easier later! In this revision aid we have tried to identify topics which are not only relevant to passing exams (a key aim of this book), but a knowledge of which will also stand you in good stead in the coming years. We have provided detailed explanations in a concise and structured format which you will find invaluable in last minute revision and which you will be able to dip into at later dates to brush up on basic concepts.

We wish you all the best for the coming exams.

We would like to thank Dr Barbara Moreland for her input in the construction of this manuscript and to our respective partners who have been supportive and tolerant and who have become excellent suppliers of hot and cold beverages and 'snacky stuff' during the wee hours. Finally, we would also like to extend our heartfelt thanks to Elliott Smock for providing us with this opportunity.

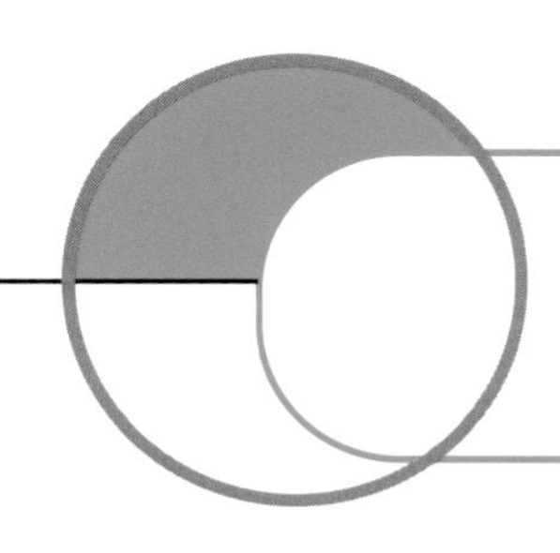

ABBREVIATIONS

ADP	adenosine diphosphate
AMP	adenosine monophosphate
AZT	azidothymine
ATP	adenosine triphosphate
bp	base pairs
cAMP	cyclic adenosine monophosphate
CAMs	cell adhesion molecules
cDNA	complementary DNA
CFTR	cystic fibrosis transmembrane conductance regulator
cGMP	cyclic guanosine monophosphate
CNS	central nervous system
COX	cyclo-oxygenase
CVS	chorionic villus sampling
DAG	diacylglycerol
dATP	deoxyadenosine triphosphate
DNA	deoxyribonucleic acid
dNTP	deoxynucleotide triphosphate
2,3-DPG	2,3-diphosphoglycerate
EFTu	elongation factor Tu
ER	endoplasmic reticulum
GDP	guanosine diphosphate
GI	gastrointestinal
GMP	guanosine monophosphate
GTP	guanosine triphosphate
HIV	human immunodeficiency virus
hnRNA	heteronuclear RNA
IP_3	inositol trisphosphate
IRS	insulin receptor substrate
LDL	low-density lipoprotein
LFT	liver function test
mRNA	messenger RNA
NMJ	neuromuscular junction
NSAIDs	non-steroidal anti-inflammatory drugs
PCNA	proliferating cell nuclear antigen
PCR	polymerase chain reaction
PNS	peripheral nervous system
RER	rough endoplasmic reticulum
RNA	ribonucleic acid
SDS-PAGE	sodium dodecyl sulphate polyacrylamide gel electrophoresis
SER	smooth endoplasmic reticulum
snRNA	small nuclear RNA
SRP	signal recognition particle
tRNA	transfer RNA
TSH	thyroid stimulating hormone
UTP	uridine triphosphate
UV	ultraviolet

SECTION 1

MEMBRANES AND LIPIDS

- CELL MEMBRANES 2
- PHOSPHOLIPIDS 4
- EICOSANOIDS 6
- THE CYCLO-OXYGENASE PATHWAYS 6
- MEMBRANE TRANSPORT (i) 8
- MEMBRANE TRANSPORT (ii) 10
- MEMBRANE PROTEINS 12
- RECEPTORS AND LIGANDS 14
- RECEPTORS AND DISEASE 16
- RECEPTOR STIMULATION 16
- SPECIALIZED MEMBRANE PROTEINS 18
- CELLULAR SIGNALLING 20
- SECOND MESSENGERS 22
- AGONISTS AND ANTAGONISTS 22
- ABDOMINAL CASE STUDY 24
- ENZYME INHIBITION 26
- ENZYME KINETICS 28
- NERVES AND ACTION POTENTIALS 30
- ACTION POTENTIALS 32
- NERVES AND MUSCLES 34

SECTION 1 MEMBRANES AND LIPIDS

1. With respect to the properties of cell membranes, which of the following statements are true?

- **a.** A lipid bilayer is composed of phospholipids with hydrophilic heads facing the outside of the membrane
- **b.** Phospholipids with unsaturated hydrocarbon tails increase membrane fluidity
- **c.** Cholesterol reduces membrane fluidity
- **d.** The lipid bilayer is approximately 50 nm thick
- **e.** Cell membranes are freely permeable to small hydrophobic molecules

2. Membrane structure and composition. From the options listed as A–J, select the most appropriate labels for points 1–5 on the diagram

Options

- **A.** Phospholipid
- **B.** Unsaturated fatty acid
- **C.** Integral protein
- **D.** Saturated fatty acid
- **E.** Glycoprotein
- **F.** Glycerol
- **G.** Glycolipid
- **H.** Surface protein
- **I.** Extracellular space
- **J.** Cholesterol

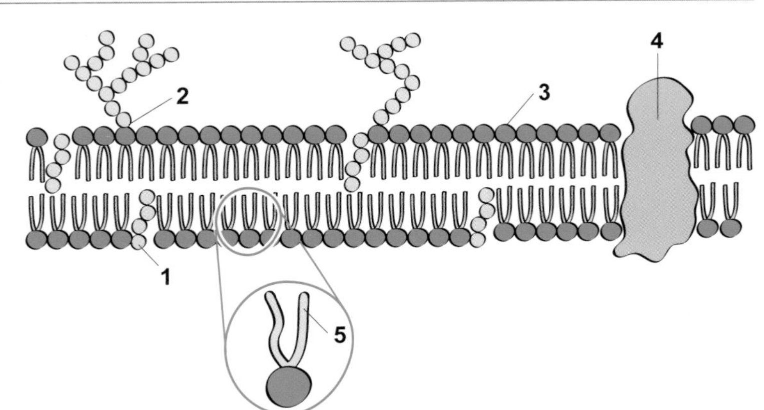

3. True or false? Phospholipid components of the cell membrane

- **a.** May freely leave the membrane
- **b.** Occasionally may flip from the extracellular to the intracellular surface
- **c.** Are amphipathic molecules
- **d.** May move freely in the plane of the membrane surface
- **e.** Sometimes have more than two fatty acid tails

EXPLANATION: CELL MEMBRANES

In water, the most energetically favourable arrangement of **phospholipids** is as a **bilayer**, with the **hydrophilic** heads facing the water, and the **hydrophobic** tails inside. Phospholipids with unsaturated hydrocarbon tails have a 'kink' at each unsaturated carbon, which makes them harder to pack tightly against the tails of their neighbours, so increasing membrane fluidity. Cholesterol, however, fills in such spaces formed by unsaturated tails and reduces fluidity. The phospholipid bilayer is approximately 5 nm thick. Small hydrophobic molecules such as O_2 and CO_2 dissolve easily in phospholipid bilayers and hence cross them freely.

The **fluid mosaic model** of the cell membrane allows phospholipids to move freely within the plane of the membrane, but the combination of hydrophilic and hydrophobic forces does not allow them to leave the membrane. **Flippases** are proteins that allow phospholipids to flip from one membrane surface to the other. This does sometimes occur even without flippases, though very rarely. An **amphipathic** molecule has both hydrophilic and hydrophobic parts, as in phospholipids and detergents. The glycerol backbone is a three-carbon unit. One carbon is esterified to the phosphate group leaving two carbons available to attach fatty acids. **Lysophospholipids** only have one fatty acid tail, but it is not possible for a phospholipid to have more than two.

Answers

1. T T T F T
2. 1 – J, 2 – G, 3 – A, 4 – C, 5 – D
3. F T T T F

4. Phospholipids

The diagram below shows a common lipid membrane component. Answer the following questions using options from the following list

Options

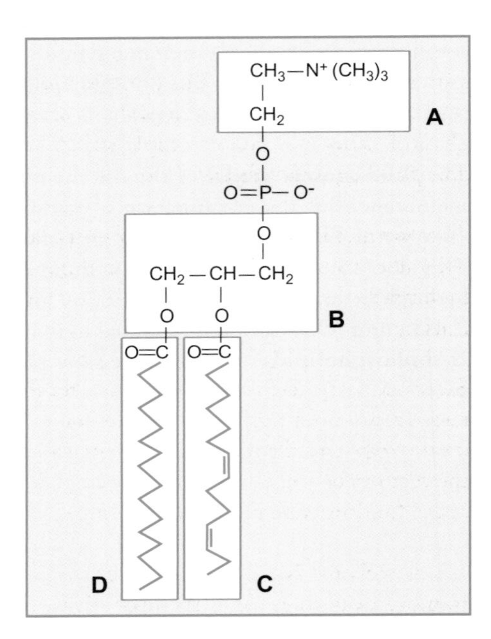

- **A.** Choline
- **B.** Glutamate
- **C.** Glutamine
- **D.** Glycerol
- **E.** Phosphate
- **F.** Phosphatidylcholine
- **G.** Phosphatidylserine
- **H.** Saturated fatty acid
- **I.** Serine
- **J.** Sphingomyelin
- **K.** Sphingosine
- **L.** Unsaturated fatty acid

1. What is the name of the whole molecule?
2. Name the group shown in box A
3. Name the group shown in box B
4. Name the group shown in box C
5. Name the group shown in box D

5. Membranes

- **a.** Briefly describe the structure of a membrane phospholipid
- **b.** Identify three factors which influence membrane fluidity and indicate whether they increase it or decrease it

6. Regarding fluidity of the plasma membrane

- **a.** Fluidity is dependent on the lipid composition of the membrane
- **b.** High concentrations of unsaturated fatty acids decrease fluidity in the phospholipid bilayer
- **c.** High concentrations of short-chain fatty acids increase fluidity
- **d.** High cholesterol concentrations decrease fluidity
- **e.** Phospholipids readily cross from one side of the membrane to the other

EXPLANATION: PHOSPHOLIPIDS

A phospholipid is composed of a glycerol molecule attached to two fatty acids and a phosphate group. The phosphate group is itself attached to a polar 'head' group. This structure renders the molecule **amphipathic**, with a hydrophilic head and a hydrophobic tail (5a).

The molecule shown is the phospholipid **phosphatidylcholine**. It contains a polar head group, choline, attached to glycerol via a phosphate ester bond. An unsaturated fatty acid is often found at the R2 position of glycerol. This molecule is amphipathic, with the choline head group being hydrophilic and the fatty acid groups being hydrophobic.

The presence of larger numbers of unsaturated fatty acids in phospholipid tails increases the membrane fluidity, and conversely higher concentrations of saturated phospholipid tails decrease fluidity. Cholesterol in the membrane decreases the fluidity. Phospholipids with short-chain fatty acids in their tails increase membrane fluidity (5b).

Saturated fatty acids are able to pack more closely than unsaturated fatty acids and because of this the presence of saturated fatty acids renders the membrane less fluid. The kinked structure of unsaturated fatty acids creates gaps and the phospholipids are less able to pack tightly together and have more freedom to move across the membrane surface. The presence of cholesterol in the membrane fills such gaps and hence decreases membrane fluidity. Membrane fluidity increases as temperature increases.

Lipids and proteins of the bilayer are free to move laterally within the membrane. However, they are rarely able to move from one side of the lipid bilayer to the other.

Answers

4. 1 – F, 2 – A, 3 – D, 4 – L, 5 – H
5. See explanation
6. T F T T F

7. Select the most appropriate answer to questions 1–5 from options A–J

Options

A. Thromboxane A_2
B. Prostaglandin E_2
C. Prothrombin
D. Aspirin
E. Leukotriene A_4
F. Cholesterol
G. Prostacyclin
H. Morphine
I. Interleukin 1B
J. Cortisol

1. Synthesized by platelets, this compound induces vasoconstriction and platelet aggregation
2. This compound is derived from arachidonic acid via the lipoxygenase pathway
3. This compound is produced by endothelial cells, inhibits platelet aggregation and promotes vasodilation
4. This compound inhibits the action of COX-1 and COX-2
5. This compound inhibits release of arachidonic acid from phospholipids by phospholipase A_2

8. Products of the cyclo-oxygenase pathways include

a. Prostacyclin
b. Cholesterol
c. Thromboxane A_2
d. Histamine
e. Prothrombin

9. The following drugs inhibit cyclo-oxygenase

a. Ibuprofen
b. Prednisolone
c. Aspirin
d. Propranolol
e. Penicillin

10. The cyclo-oxygenase pathways

a. Name a single enzyme which separates arachidonic acid from membrane phospholipids
b. The COX pathways process arachidonic acid – name one other pathway
c. What is the essential difference between the COX-1 and COX-2 pathways?
d. Name three products of the COX pathways
e. For two of the products you named in (d) give an example of their local effects
f. Name two drugs which inhibit the COX pathways
g. Recently, anti-inflammatory drugs have been developed which selectively inhibit one of the COX pathways. Which of the pathways do they inhibit, and why is such selective inhibition desirable?

COX, cyclo-oxygenase; NSAIDs, non-steroidal anti-inflammatory drugs

EXPLANATION: EICOSANOIDS

The **eicosanoids** are a group of compounds that are synthesized mostly from arachidonic acid, but also from derivatives of linoleic and linolenic acids. The three main enzyme pathways acting on arachidonic acid are the **cyclo-oxygenase**, lipoxygenase and cytochrome P450 mono-oxygenase pathways. The **cyclo-oxygenase pathway** leads to production of prostaglandins, prostacyclin and thromboxanes. The lipoxygenase pathway produces lipoxins and leukotrienes. The production of eicosanoids can be prevented at a number of points in their synthesis pathways. NSAIDs inhibit the action of cyclo-oxygenase, whereas steroids such as endogenous cortisol inhibit liberation of **arachidonic acid** from phospholipids by phospholipase A_2.

The prostanoids are a group of compounds produced by the COX pathways. They include prostacyclin, thromboxane A_2 and the prostaglandins. Cholesterol is synthesized by a separate pathway involving combination of multiple two-carbon units. Histamine is a basic amine synthesized from histidine.

EXPLANATION: THE CYCLO-OXYGENASE PATHWAYS

Phospholipase A_2 releases **arachidonic acid** from phospholipids in the plasma membrane (10a). Arachidonic acid can also be liberated by phospholipase C and diacylglycerol lipase, or phospholipase D and phospholipase A_2. The lipoxygenase and cytochrome P450 mono-oxygenase pathways both produce **eicosanoids** from arachidonic acid (10b).

The **COX-1** pathway is involved in continuous physiological processes, whereas the **COX-2** pathway is an induced pathway and only operates under certain circumstances (10c). Both pathways produce prostaglandins, prostacyclin and thromboxane A_2 (10d).

Prostaglandins have a very wide range of actions. Some are involved in regulation of gastric acid secretion, while some are able to induce spontaneous abortion or labour in pregnant women. Prostaglandins also stimulate renin secretion, and may have a role in regulating plasma free fatty acid levels. **Prostacyclin** inhibits platelet aggregation and has a vasodilating effect, whereas **thromboxane A_2** promotes platelet aggregation and is a vasoconstrictor (10e).

NSAIDs inhibit the **COX pathways**. Examples are aspirin, ibuprofen, indomethacin and diclofenac (10f).

Drugs which selectively block the **COX-2 pathway** have been developed. These do not interfere with the constitutive **COX-1 pathway**, and hence are useful in people with clotting disorders or gastric ulceration in whom normal NSAIDs are contraindicated (10g).

Answers

7. 1 – A, 2 – E, 3 – G, 4 – D, 5 – J
8. T F T F F
9. T F T F F
10. See explanation

11. Regarding transporters of biological membranes

a. Uniporters move a single substance across a membrane
b. Symporters move two different substances simultaneously in opposite directions across the membrane
c. The Na^+/K^+ ATPase is an example of an antiporter
d. The glucose transporter GLUT 1 is a symporter
e. The intestinal Na^+/glucose symporter is an example of a secondary active transporter

12. With regard to passive diffusion across a cell membrane

a. Passive diffusion requires ATP expenditure
b. The rate of flow is proportional to the concentration gradient
c. The rate of flow is inversely proportional to the hydrostatic pressure gradient
d. The rate is higher for more hydrophilic molecules
e. The flow requires specialized membrane-bound proteins

13. Passive transport across biological membranes

a. Name and describe two mechanisms of passive transport across the plasma membrane
b. Briefly describe the kinetics of transport for each mechanism
c. Describe the property of the substances transported by the mechanisms and give two examples for each mechanism

ATP, adenosine triphosphate

EXPLANATION: MEMBRANE TRANSPORT (i)

Membrane transporters can be classified as **uniporters**, **symporters** or **antiporters**. **Uniporters** move a single substance, such as glucose, across a membrane one molecule at a time. **Symporters** simultaneously transport two different substances in **the same direction** across the membrane. Several amino acids and glucose are transported in this way, with Na^+ being co-transported (e.g. the intestinal Na^+/glucose transporter, which is also an example of secondary active transport). **Antiporters** transport two different substances simultaneously in opposite directions across the membrane (e.g. the Na^+/K^+ ATPase pump).

In **simple diffusion**, substances pass directly through the plasma membrane, down their concentration gradient. In **facilitated diffusion**, transport across the membrane requires a pore or a carrier protein situated within the bilipid layer. Substances binding to the carrier protein on one side of the membrane cause a conformational change in the protein that results in the transfer of the substance to the other side. Diffusion is down the concentration gradient and does not require the hydrolysis of ATP (13a).

The rate of transport of substances by **simple diffusion** is directly related to the concentration gradient and the lipid solubility of the substance, i.e. the rate of transport will increase linearly with increasing concentration gradients and lipid solubility, and there is no theoretical maximal rate of transport. In **facilitated diffusion**, however, the rate of transport is limited by the number of carrier molecules in the plasma membrane. As the concentration of the substance increases for a given amount of carrier molecule, there will come a point when all the carrier proteins are occupied. Hence further increases in substance concentration or the concentration gradient will not increase the rate of transport, the mechanism is **saturable** (13b).

Molecules that cross the membrane by **simple diffusion** are either hydrophobic (e.g. steroid hormones, oxygen, benzene and short-chain fatty acids) or are small uncharged polar molecules (e.g. water, carbon dioxide and urea). Charged and large polar molecules cannot pass in this way. Such molecules can, however, be transported by **facilitated diffusion** using carrier proteins or pores (e.g. glucose, sucrose, Na^+ ions, Cl^- ions) **(13c)**.

Answers

11. T F T F T
12. F T F F F
13. See explanation

14. Facilitated diffusion of a substance across a membrane

- **a.** Is an example of passive transport
- **b.** Requires a pore or carrier protein
- **c.** Only occurs down the electrochemical gradient of the substance
- **d.** Requires energy derived from the hydrolysis of ATP or GTP
- **e.** Is the mechanism by which most steroid hormones enter the cell

15. True or false? Facilitated diffusion across a cell membrane

- **a.** Requires ATP expenditure
- **b.** Requires the action of specialized proteins
- **c.** Occurs against the concentration gradient
- **d.** Has a maximum rate at which diffusion can take place
- **e.** Occurs at a rate that is higher for more hydrophilic molecules

16. Active transport (across a biological membrane)

- **a.** Allows transfer of substances against an electrochemical gradient
- **b.** Is required to maintain the plasma membrane potential of cells
- **c.** Is said to be primary when the transporter protein hydrolyses ATP
- **d.** Is said to be secondary when the transporter protein hydrolyses GTP
- **e.** Of H^+ ions into the stomach lumen by the H^+ ion pump of parietal cells is an example of secondary active transport

17. Active transport across a cell membrane

- **a.** Obeys Michaelis–Menten kinetics
- **b.** Requires ATP expenditure
- **c.** Can occur against the concentration gradient
- **d.** Can only transport large molecules
- **e.** Requires specialized transport proteins

ATP, adenosine triphosphate; GTP, guanosine triphosphate

EXPLANATION: MEMBRANE TRANSPORT (ii)

Passive transport of substances across biological membranes occurs by **simple diffusion** or by **facilitated diffusion**. Energy from the hydrolysis of ATP or GTP is not required for these processes. The energy required for the movement of these substances is derived from the electrochemical gradient of the substance across the membrane; substances move from areas of high concentration to areas of lower concentration. **Facilitated diffusion** is mediated by pores or carrier proteins which help move the substance across the membrane. These mechanisms are able to mediate transport in either direction. Steroid hormones, being lipophilic, are able to traverse the lipid membrane by **simple diffusion** and do not require a facilitated mechanism.

The process of facilitated diffusion is saturable – there is a maximum rate at which transport can take place. The rate of transfer does not depend on the hydrophilicity of the substrate alone and depends largely on the properties of the carrier protein.

Active transport of substances requires energy that is ultimately derived from the hydrolysis of nucleotide triphosphates such as ATP or GTP. Such transport systems can be described as **primary** or **secondary active transport mechanisms**. In **primary transport mechanisms**, the transporter protein directly hydrolyses ATP during the transfer. An example of this is the Na^+/K^+ ATPase, which is largely responsible for maintaining the plasma membrane potential of the cell. Another example is the H^+ ion pump of parietal cells, where transport of H^+ ions into the stomach lumen is mediated by hydrolysis of ATP. In **secondary active transport**, the transporter does not directly hydrolyse ATP. Instead, it uses the energy stored in the electrochemical gradient of other substances (usually Na^+) to drive transport. The Na^+ gradient is maintained by the Na^+/K^+ ATPase pump by hydrolysis of ATP. The process requires transport proteins and obeys **Michaelis–Menten kinetics**.

Answers

14. T T T F F
15. F T F T F
16. T T T F F
17. T T T F T

18. Membrane proteins

The diagram below shows the plasma membrane of a cell. The shapes indicate proteins that are attached to the membrane. For each of the questions, select a phrase from the list A–J, which is most descriptive

Options

- **A.** Protein 1
- **B.** Protein 2
- **C.** Protein 3
- **D.** Protein 4
- **E.** Bitopic integral membrane protein
- **F.** G protein-coupled receptor
- **G.** Lipid-anchored protein
- **H.** Monotopic integral membrane protein
- **I.** Peripheral membrane protein
- **J.** Polytopic integral membrane protein

1. What sort of membrane protein is 1?
2. What sort of membrane protein is 2?
3. What sort of membrane protein is 3?
4. What sort of membrane protein is 4?
5. Which protein in the above diagram can be removed from the cell by washing with a solution containing a high salt concentration?

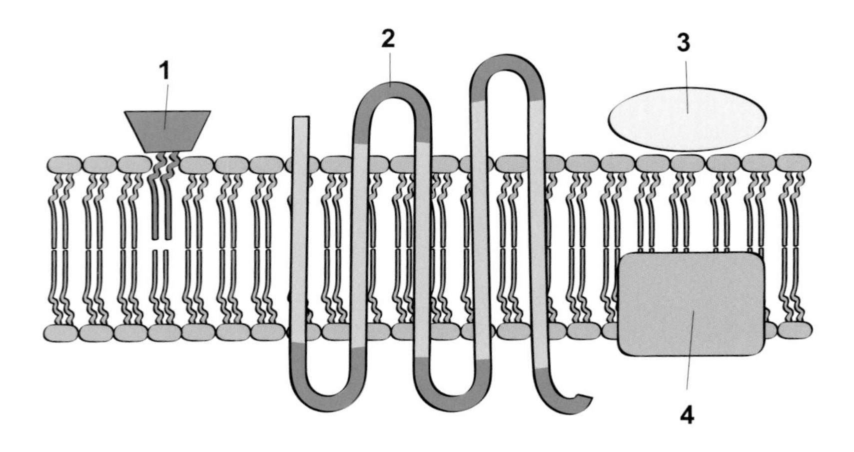

EXPLANATION: MEMBRANE PROTEINS

Membrane proteins associate with the membrane in a number of different ways, appropriate to their function.

Protein 1 in the diagram opposite is a **lipid-anchored protein**. The protein contains a covalently bound lipid anchor which inserts into the plasma membrane. The protein may be removed from the membrane by the action of phospholipases which will cleave the lipid anchor.

Protein 2 is a **polytopic integral membrane protein**. It is securely inserted into the plasma membrane and usually can only be removed by harsh treatments (e.g. detergents).

Protein 3 is a **peripheral membrane protein**, which interacts relatively weakly with the membrane. It can be removed easily with gentle treatments such as washing the cells in a buffer containing high salt concentration or a solution containing urea.

Protein 4 is a **monotypic integral membrane protein**. It does not pass through both layers of the bilayer lipid membrane.

Answers

18. 1 – G, 2 – J, 3 – I, 4 – H, 5 – C

19. Low-density lipoproteins

The following questions concern the low-density lipoprotein (LDL) receptor and its ligand. Choose the most descriptive/appropriate phrase from the list below for each question

Options

A. Apolipoprotein E_2
B. Apoprotein B100
C. Clathrin-coated pits
D. Cholesterol
E. Cytoplasm
F. Familial hypercholesterolaemia
G. Heterozygotes
H. Heterozygotes and homozygotes
I. Homozygotes
J. Hyperalbuminaemia
K. Lysosomal storage disease
L. Lysosome
M. Plasma membrane

1. Where is the LDL receptor localized within a cell?
2. What is the ligand for the LDL receptor?
3. What does the LDL ligand transport?
4. Defects in the gene for the receptor result in which disease state?
5. In which genotype does the disease occur?

20. Regarding the nicotinic acetylcholine receptor

a. It is composed of three alpha- and two beta-subunits
b. Acetylcholine binds to the alpha-subunits
c. Two acetylcholine molecules are required to maximally open the channel
d. It does not require ATP
e. It facilitates the transport of Na^+ ions against a concentration gradient

21. Regarding the insulin receptor

a. It is an example of a G protein-coupled receptor
b. It is a member of the family of glucose transporters
c. Insulin binds to the alpha-subunits
d. Insulin binding results in the autophosphorylation of the beta-subunits
e. Glucagon is a reversible competitive antagonist of the receptor

ATP, adenosine triphosphate; IRS, insulin receptor substrate; LDL, low-density lipoprotein

EXPLANATION: RECEPTORS AND LIGANDS

LDL transports cholesterol in the plasma. It is the principal mechanism of cholesterol transport in humans. The **apoprotein** B100 component of LDL is the ligand for the LDL receptor. The receptor is synthesized in the **endoplasmic reticulum**, processed in the **Golgi apparatus** and is transported to the plasma membrane where it localizes to clathrin-coated pits. Defects in the receptor result in **familial hypercholesterolaemia**. The genetic lesion is **autosomal dominant** and hence heterozygotes and homozygotes are affected. The disease presents earlier and is much more severe in homozygotes.

The **nicotinic acetylcholine receptor** is an ion channel composed of five subunits (two alpha, one beta, one gamma and one delta). Binding of an **acetylcholine** molecule to each of the alpha-subunits induces a conformational change that results in the channel becoming fully open. When the channel is open, small cations such as Na^+ move into the cell down a concentration gradient, thus depolarizing the cell membrane. This process does not require the hydrolysis of ATP.

The **insulin receptor** is composed of two extracellular-facing ligand binding alpha-subunits which bind the ligand and two intracellular-facing beta-subunits. It is an example of a **tyrosine kinase receptor** and binding of insulin induces a conformational change that activates the intracellular tyrosine kinase (beta subunit). This results in the phosphorylation of tyrosine residues on the beta-subunit (regulation of the receptor) and of the insulin receptor substrate (IRS). Phosphorylation of the IRS ultimately leads to the increase of glucose transporters at the cell surface that facilitate the sequestration of glucose.

Answers

19. 1 – C, 2 – B, 3 – D, 4 – F, 5 – H
20. F T T T F
21. F F T T F

22. Receptor dysfunction

Choose one disease from the list below (A–L) which results from defects in the receptor proteins indicated in each of the questions

Options

A. Acromegaly
B. Familial hypercholesterolaemia
C. Gilbert's syndrome
D. Leprechaunism
E. Osteogenesis imperfecta
F. Retinitis pigmentosa
G. Tay–Sachs disease
H. Testicular feminization
I. Type I diabetes mellitus
J. Type I rickets
K. Type II rickets
L. X-linked colour blindness

1. Androgen receptor
2. Vitamin D receptor
3. Rhodopsin
4. LDL receptor
5. Insulin receptor

23. True or False? The following diseases are caused by autoantibodies to cell surface receptors

a. Cystic fibrosis
b. Graves' disease
c. Type II diabetes mellitus
d. Myasthenia gravis
e. Testicular feminization syndrome

24. Cell surface receptors

Pick the most appropriate outcome from options A–K on stimulation of the cell surface receptors listed below

Options

A. Decrease in intracellular cyclic AMP
B. Decrease in intracellular cyclic GMP
C. Efflux of Na^+ ions from the cell
D. Increase in intracellular arachidonic acid
E. Increase in intracellular ceramide
F. Increase in intracellular cyclic AMP
G. Increase in intracellular cyclic GMP
H. Increase in intracellular inositol triphosphate
I. Influx of Na^+ ions into the cell
J. Phosphorylation of serine residues
K. Phosphorylation of tyrosine residues

1. Insulin receptor
2. Nicotinic acetylcholine receptor
3. Muscarinic acetylcholine receptor
4. Beta adrenergic receptor
5. Alpha-2 adrenergic receptor

TSH, thyroid stimulating hormone; LDL, low-density lipoprotein; AMP, adenosine monophosphate; GMP, guanosine monophosphate; DAG, diacylglycerol; IP_3, inositol trisphosphate

EXPLANATION: RECEPTORS AND DISEASE

When a mutation occurs in an androgen receptor, **androgen insensitivity** may occur. The **genotype** is male (XY), but the **phenotype** is outwardly female. Although testosterone is produce by the testes, there are no receptors able to respond to it and development of the embryo proceeds as a female. However, a uterus is not present and the vagina is blind ending.

Defects in the rhodopsin light receptors can cause **retinitis pigmentosa**. This is an **autosomal dominant** condition which initially causes loss of peripheral vision, progressing to complete loss of vision.

Leprechaunism is a rare and fatal disease caused by a defect of the insulin receptor. There is severe insulin resistance and the syndrome is characterized by 'elfin' like facies, decreased subcutaneous fat, hirsutism and intrauterine and neonatal growth retardation.

Cell surface receptors can be the targets of antibodies produced in autoimmune diseases. Two notable examples are **Graves' disease** and **myasthenia gravis**. In Graves' disease, antibodies are produced against the TSH receptor found on the follicular cells of the thyroid. These antibodies stimulate the TSH receptor, thus causing synthesis and release of thyroxine in the absence of pituitary TSH; the net result is hyperthyroidism. Graves' disease is the commonest cause of hyperthyroidism.

Antibodies to muscle cell surface nicotinic acetylcholine receptors are produced in **myasthenia gravis**. The result is the inactivation of these receptors, thus causing muscle weakness.

EXPLANATION: RECEPTOR STIMULATION

There are a vast number of cell surface receptors. The effects of ligand–receptor binding of a few of the most commonly encountered receptors are presented here:

Ligand type	Receptor type	Effect
Insulin receptor	Tyrosine kinase receptor	Phosphorylation of tyrosine residues on target proteins and on the insulin receptor itself.
Nicotinic acetylcholine receptor	Ion channel	Na^+ ions influx
Muscarinic receptor	G protein-coupled receptor	Activates phospholipase C resulting in the production of DAG and IP_3
Beta adrenergic receptors	G protein-coupled receptor	Activates adenylyl cyclase on stimulation. The result is increase of cyclic AMP intracellularly
Alpha-2 adrenergic receptor	G protein-coupled receptor	Inhibits the activity of adenylyl cyclase. The net result of stimulating this receptor is a decrease in intracellular cyclic AMP

Answers

22. 1 – H, 2 – K, 3 – F, 4 – B, 5 – D
23. F T F T F
24. 1 – K, 2 – I, 3 – H, 4 – F, 5 – A

25. The following drugs act predominantly on the indicated cell surface receptor

- **a.** Metoprolol: alpha adrenergic receptors
- **b.** Isoprenaline: beta adrenergic receptors
- **c.** Atracurium: acetylcholine receptors
- **d.** Atropine: muscarinic acetylcholine receptors
- **e.** Neostigmine: muscarinic acetylcholine receptors

26. Regarding the Na^+/K^+ ATPase pump

- **a.** It is an example of primary active transport
- **b.** Three Na^+ ions are exchanged out of the cell for every two K^+ ions transported inside
- **c.** Binding of K^+ ions to the transporter causes autophosphorylation
- **d.** Binding of Na^+ ions directly causes a conformational change in the transporter
- **e.** Digoxin inhibits the function of the pump

27. The following receptors have an intrinsic enzyme activity

- **a.** Insulin receptor
- **b.** Natriuretic peptide receptor
- **c.** Glucagon receptor
- **d.** Alpha adrenergic receptor
- **e.** Nicotinic acetylcholine receptor

28. True or false? Tyrosine kinase receptors

- **a.** Have no membrane spanning domains
- **b.** Have a tyrosine kinase activity that is located within the plasma membrane
- **c.** Can become autophosphorylated
- **d.** Have an extracellular domain that binds to ligand
- **e.** Once bound to ligand move into the nucleus, where they interact with the regulatory elements of genes

ATP, adenosine triphosphate; GTP, guanosine triphosphate; GMP, guanosine monophosphate

EXPLANATION: SPECIALIZED MEMBRANE PROTEINS

Metoprolol is a **beta-blocker**, i.e. it selectively blocks the beta adrenergic receptors. It is used in the treatment of hypertension, angina pectoris and in migraine. Isoprenaline is a **sympathomimetic** that acts almost exclusively on beta adrenergic receptors. Atracurium is a competitive inhibitor of **acetylcholine receptors** and is used to induce muscle paralysis during operative procedures. Neostigmine is an **acetylcholinesterase antagonist** and can be used to reverse the effects of atracurium. Atropine is a muscarinic selective antagonist and can be used in cases of bradycardia and during anaesthesia to prevent secretions within the respiratory tract.

The **Na^+/K^+ ATPase pump** transports three Na^+ ions (intracellular to extracellular) and two K^+ ions (extracellular to intracellular) across the plasma membrane against a concentration gradient. The energy required is derived from the hydrolysis of ATP (primary active transport). Binding of Na^+ (not K^+) to the pump induces autophosphorylation by hydrolysis of ATP. The **phosphorylation** of the receptor induces a **conformational change** that facilitates transport. Plant glycosides such as digoxin inhibit this process by binding to the K^+ binding site.

The **insulin receptor** has an intrinsic **tyrosine kinase** activity that is activated by binding of insulin. Similarly, the natriuretic peptide receptor has a guanylate cyclase activity (GTP is converted to cyclic GMP). The glucagon and alpha adrenergic receptors are **G protein-coupled receptors** and do not have an intrinsic enzyme activity.

Tyrosine kinase receptors (e.g. the insulin receptor), are membrane spanning proteins (often multimeric in structure), which have a ligand binding domain located extracellularly and a catalytic domain situated intracellularly. Binding of the ligand causes a conformational change that activates the tyrosine kinase activity. This activity phosphorylates tyrosine residues of target proteins and tyrosine residues of the receptor itself. Receptor autophosphorylation regulates the activity of the receptor.

Answers

25. F T T T F
26. T T F F T
27. T T F F F
28. F F T T F

29. True or false? The following substances can act as secondary messengers within cells

a. Na^+ ions
b. Cyclic AMP
c. Cyclic GMP
d. Ceramide
e. Diacylglycerol

30. G protein-coupled receptors

a. Have seven transmembrane spanning domains
b. Have an N-terminus that is glycosylated
c. Have a large intracellular loop that interacts with G proteins
d. When activated by ligand, bind a molecule of GTP
e. When activated, cause an increase in intracellular cyclic GMP concentration

31. Activation of phospholipase C

a. Leads to the hydrolysis of phosphatidylinositol bisphosphate
b. Causes a rise in intracellular inositol triphosphate concentration
c. Causes a reduction in intracellular diacylglycerol concentration
d. Causes a reduction in cytosolic free Ca^{2+} concentration
e. Causes deactivation of protein kinase C

32. Protein kinase C is activated directly by

a. Diacylglycerol
b. Ca^{2+} ions
c. Cyclic AMP
d. Inositol triphosphate
e. Ceramide

DAG, diacylglycerol; GTP, guanosine triphosphate; GDP, guanosine diphosphate; GMP, guanosine monophosphate; IP_3, inositol trisphosphate; AMP, adenosine monophosphate

EXPLANATION: CELLULAR SIGNALLING

Secondary messengers mediate the transduction of signals from an activated receptor to other proteins situated within the cytoplasm of the cell. Many molecules can act as secondary messengers within the cell, including **cyclic AMP**, **cyclic GMP**, **ceramide** and **DAG**.

G protein-coupled receptors in general have **seven transmembrane spanning domains** and have their N-terminals (which are glycosylated) located outside the cell and their C-terminals located intracellularly. When bound to ligand, these receptors interact with G proteins via a large intracellular loop (usually between the fifth and sixth transmembrane domains). GTP binds to the G protein complex when the complex interacts with the receptor; the receptor does not directly bind GTP. On GTP binding, the alpha-subunit of the G protein complex disassociates. As the complex reforms, the GTP is hydrolysed to GDP. Cyclic GMP is not formed.

Phospholipase C hydrolyses phosphotidylinositol bisphosphate (a minor component of the plasma membrane) into IP_3 and DAG. IP_3 binds to receptors on the endoplasmic reticulum and causes a release of free Ca^{2+} into the cytosol from intracellular stores. Both Ca^{2+} and DAG activate protein kinase C. IP_3 therefore does not directly stimulate protein kinase C; however, it mediates the release of Ca^{2+} ions into the cytoplasm from intracellular stores, thus activating protein kinase C indirectly. Cyclic AMP activates protein kinase A.

Answers

29. F T T T T
30. T T T F F
31. T T F F F
32. T T F F F

33. Briefly explain the process that leads to the production of the second messenger cyclic AMP upon stimulation of the beta adrenergic receptor by adrenaline

34. Concerning second messengers

From the options below pick the enzyme that produces the second messengers listed as 1–5

Options

A. Adenylyl cyclase
B. Cerebrosides
C. Creatine kinase
D. Cyclo-oxygenase II
E. Guanylyl cyclase
F. Nitric oxide synthase
G. Phospholipase A_2
H. Phospholipase C
I. Phospholipase D
J. Protein kinase C
K. Protein tyrosine phosphatase
L. Sphingomyelinase

1. Cyclic AMP
2. Ceramide
3. Arachidonic acid
4. Diacylglycerol
5. Phosphatidic acid

35. True or false? Regarding receptor agonists

a. Agonist binding activates receptors by inducing a conformational change
b. An increase in agonist concentration always produces a proportional increase in receptor response
c. Partial agonists have high efficacy though they elicit a lower receptor response than full agonists
d. Increasing the dose of a partial agonist will eventually lead to a maximal receptor response
e. Acetylcholine is an agonist of muscarinic receptors

36. Regarding receptor antagonists

a. The effect of irreversible competitive antagonists on receptors is surmountable by increasing the agonist concentration
b. The muscle relaxant pancuronium, which is used during surgery, is an irreversible competitive antagonist of acetylcholine receptors
c. Non-competitive antagonists exert their effect by binding irreversibly to the agonist binding site
d. Competitive antagonists occupy the same receptor binding site as agonists
e. Reversible competitive antagonists are useful clinically

AMP, adenosine monophosphate; ATP, adenosine triphosphate; DAG, diacylglycerol; GTP, guanosine triphosphate; GDP, guanosine diphosphate; GMP, guanosine monophosphate

EXPLANATION: SECOND MESSENGERS

The beta adrenergic receptor is a transmembrane **G protein-coupled receptor**. Binding of the ligand (adrenaline) causes a conformational change in the receptor that allows attachment to the G protein complex. The resting G protein complex is composed of three proteins (stimulatory alpha, beta and gamma) and is located in the plasma membrane. A molecule of GDP is found attached to the alpha-subunit. Upon binding to the receptor, the GDP is released and a GTP molecule becomes bound to the alpha-subunit. This causes disassociation of the G protein complex in the plasma membrane. The stimulatory alpha-subunit binds to adenylyl cyclase (adenylate cyclase), which becomes activated. **Adenylyl cyclase** hydrolyses ATP to cyclic AMP, the second messenger (33). An intrinsic GTPase activity cleaves the GTP attached to the alpha-subunit and the G protein complex reforms.

Some common enzymes and their second messengers are summarized in the following table:

Second messenger	Mechanism/substrate	Enzyme
Cyclic AMP	Via the hydrolysis of ATP	Adenylyl cyclase
Ceramide	Sphingomyelin	Sphingomyelinase
Arachidonic acid	Plasma membrane phospholipids	Phospholipase A_2
DAG	Plasma membrane phospholipids	Phospholipase C
Phosphatidic acid	Plasma membrane phospholipids	Phospholipase D

EXPLANATION: AGONISTS AND ANTAGONISTS

As indicated in the diagram opposite, receptor response does not increase proportionally with agonist concentration.

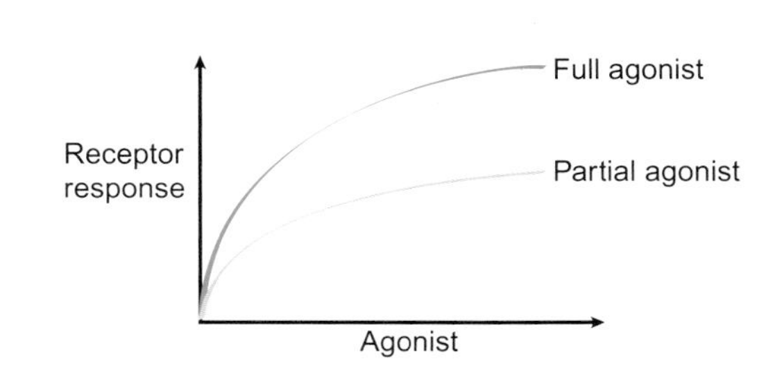

Efficacy describes the ability of an agonist to activate the receptor. Partial agonists are limited in their ability to activate receptors and thus have low efficacy. Increasing the concentration of partial agonists will not result in a maximal receptor response as shown in the diagram.

Competitive antagonists bind to the same site as agonists. They may be reversible or irreversible. Whilst the effects of **reversible competitive antagonists** can be overcome by increasing the agonist concentration, those of **irreversible competitive antagonists** cannot; they bind irreversibly to the agonist binding site. Reversible competitive antagonists such as pancuronium and propanolol are clinically useful, whilst irreversible competitive antagonists are generally not. Non-competitive antagonists do not compete for the agonist binding site. They bind elsewhere on receptors.

Answers

33. See explanation
34. 1 – A, 2 – L, 3 – G, 4 – H, 5 – I
35. T F F F T
36. F F F T T

37. Case study

A 74-year-old man is to undergo a major laparotomy for the removal of a cancerous growth of the colon. As the surgeon makes the initial incision, she finds that the abdominal muscles are tense and realizes that the surgery might be difficult. She asks the anaesthetist if anything can be done to relax the abdominal muscles.

- **a.** Suggest how the anaesthetist may help. Briefly discuss the mode of action of drugs that may be used
- **b.** What complications arise from the use of such drugs?
- **c.** The surgery is completed successfully. How can the anaesthetist reverse the effects of the administered drug?

NMJ, neuromuscular junction

EXPLANATION: ABDOMINAL CASE STUDY

The terminals of neurons innervating the abdominal muscles release acetylcholine, which crosses the synapse and stimulates **nicotinic acetylcholine receptors** on the post-synaptic muscle cell membrane. These nicotinic acetylcholine receptors are ligand-gated ion channels which allow entry of Na^+ into the cells on activation. The entry of Na^+ ions causes a depolarization of the muscle cell membrane and this eventually leads to muscle contraction. Thus muscle contraction may be inhibited by using a competitive drug that blocks the action of acetylcholine on its receptor. Drugs that selectively block nicotinic acetylcholine receptors at the neuromuscular junction are called neuromuscular blocking agents and include:

- Vecuronium
- Atracurium
- Pancuronium
- Rocuronium
- Mivacurium

Such drugs are often used to relax musculature during surgery, especially when the incision traverses skeletal muscle. Each type of drug has a different duration of action (e.g. mivacurium is a short-acting agent, atracurium is an intermediate-acting agent and pancuronium is a long-acting agent) (37a).

These agents will also paralyse the diaphragm and hence the patient needs to be intubated and ventilated. Furthermore, these agents should only be administered to patients who are unconscious as the generalized muscle paralysis they induce would cause much distress (37b).

Because the drugs described above are competitive inhibitors, increase in the available acetylcholine will overcome their effects. Normally, the acetylcholine in the NMJ is broken down by the action of acetylcholinesterase. Therefore, inhibition of this enzyme will lead to higher concentrations of acetylcholine in the NMJ and hence reverse the paralysis. Reversible inhibitors of acetylcholinesterase include:

- Neostigmine
- Edrophonium
- Pyridostigmine

These drugs will also increase the acetylcholine concentrations at muscarinic receptors and this may cause **muscarinic effects** such as **increased bronchial secretions**, **bronchospasm**, **bradycardia** and **intestinal hypermotility**. Hence to overcome these side effects, these drugs are usually given with an antimuscarinic agent such as glycopyrrolate (37c).

Answers

37. See explanation

38. True or false? Competitive reversible enzyme inhibitors

- **a.** Can bind to the substrate in a reversible fashion
- **b.** Form covalent bonds with the enzyme–substrate complex
- **c.** Often have a similar structure to the substrate of the enzyme
- **d.** Increase the K_m (as described by the substrate concentration required for half-maximal enzyme activity)
- **e.** Decrease V_{max} (maximum enzyme velocity rate)

39. Irreversible enzyme inhibitors

- **a.** Form covalent bonds with the enzyme or enzyme–substrate complex
- **b.** Decrease the concentration of active enzyme in a system
- **c.** Do not alter K_m
- **d.** Do not alter V_{max}
- **e.** Are exemplified by cyanide and cytochrome oxidase

40. Regarding enzyme inhibition. Use the suggestions to complete the sentences below

Options

- **A.** Allosteric inhibitors
- **B.** Reversible inhibitors
- **C.** Irreversible inhibitors
- **D.** Increased
- **E.** Decreased
- **F.** Unchanged
- **G.** Competitive inhibitors
- **H.** Non-competitive inhibitors

1. In competitive inhibition of an enzyme, V_{max} is
2. In competitive inhibition of an enzyme, K_m is
3. In non-competitive inhibition of an enzyme, V_{max} is
4. Bind to enzymes non-covalently and can be removed by dialysis
5. Bind at the same site on the enzyme as the substrate

HIV, human immunodeficiency virus; AZT, azidothymidine

EXPLANATION: ENZYME INHIBITION

Competitive reversible enzyme inhibitors bind to the active site of the enzyme; they do not bind to the substrate. They occupy the same site as the enzyme substrate and hence often have a structure very similar to that of the substrate. The binding of the inhibitor is reversible and covalent bonds are not formed. The effects of these types of inhibitors can be overcome by increasing the substrate concentration and therefore the maximum velocity of the catalysis remains unchanged. However, a higher substrate concentration is required to attain any given reaction rate in the presence of these inhibitors and hence K_m is increased.

Irreversible enzyme inhibitors covalently bind to either the enzyme or enzyme–substrate complex, thus inactivating the enzyme; consequently the concentration of active enzyme decreases. As the amount of active enzyme in a system is reduced the maximal velocity (V_{max}) is also reduced. The K_m, however, is unaltered. Cyanide is a classic example of an irreversible inhibitor; it covalently binds to mitochondrial cytochrome oxidase and inhibits all reactions associated with electron transport.

Modulation of enzyme function is the action of many drugs commonly used in clinical practice. Enzymes are often the targets for naturally occurring and also manufactured poisons and inhibitors. Neostigmine inhibits the action of acetylcholinesterase in the neuromuscular junction. Acetylcholinesterase breaks down acetylcholine and is irreversibly inhibited by the nerve gas sarin. AZT is a drug used in the treatment of HIV which inhibits viral reverse transcriptase. HIV is a retrovirus that uses the enzyme reverse transcriptase to produce DNA from its own RNA genome.

Answers

38. F F T T F
39. T T T F T
40. 1 – F, 2 – D, 3 – E, 4 – B, 5 – G

41. Enzyme kinetics

The diagram below shows Lineweaver–Burk plots of a reaction catalysed by an enzyme. The solid line shows the kinetics of the uninhibited enzyme. The dashed lines show kinetics of the same enzyme with various types of inhibitors. Pick an appropriate answer from the list (A–M) for each of the questions

Options

A. Point A
B. Point B
C. Curve C
D. Curve D
E. Curve E
F. 1/A
G. −1/A
H. 1/B
I. −1/B
J. 1/substrate concentration
K. 1/initial velocity
L. Substrate concentration
M. Velocity

1. Label the *x* axis
2. Label the *y* axis
3. The K_m is equivalent to
4. The V_{max} is equivalent to
5. The addition of a competitive reversible inhibitor to the enzyme reaction will give rise to the curve labelled

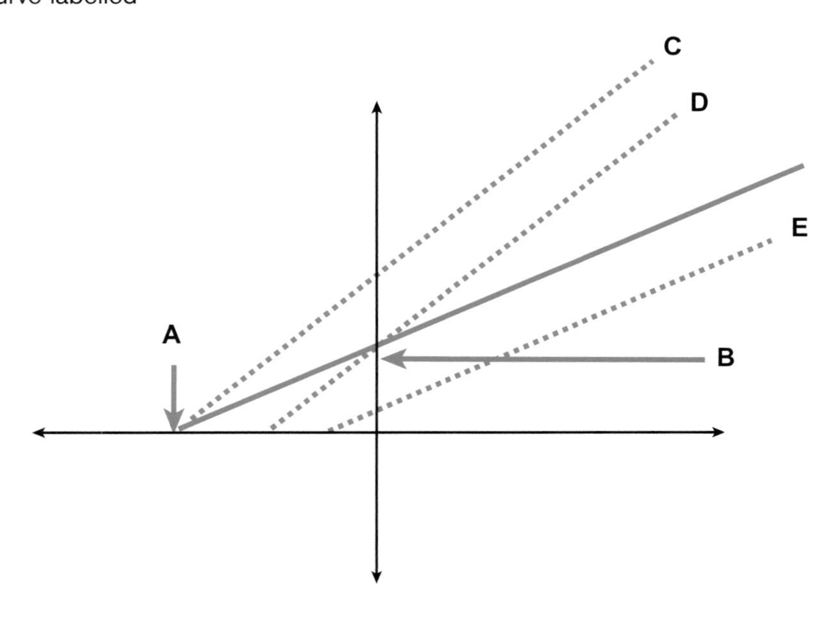

EXPLANATION: ENZYME KINETICS

The **Lineweaver–Burk plot** is derived from the equation below (a rearrangement of the **Michaelis–Menten equation**):

$$1/V_o = K_m/(V_{max} \cdot [S]) + 1/V_{max}$$

and 1/substrate concentration (x axis) is plotted against 1/velocity (y axis).

The Lineweaver–Burk plot is a more accurate and linear way of expressing the Michaelis–Menten equation. By taking the reciprocal of both sides of the Michaelis–Menten equation, a straight line graph of $1/V_0$ against $1/[S]$ can be plotted. The Lineweaver–Burk plot gives the equation of a straight line ($y = mx + c$) where $y = 1/V_0$, $m = K_m/V_{max}$, $x = 1/[S]$, $c = 1/V_{max}$. This line crosses the vertical ($1/V_0$) axis at $1/V_{max}$ and the horizontal ($1/[S]$) axis at $-1/K_m$. Values of K_m and V_{max} are more accurately measured from the Lineweaver–Burk plot than from a plot of the Michaelis–Menten equation.

The intersection of the x axis is equal to $-1/K_m$ and therefore the K_m is equal to $-1/A$ in question 41. The V_{max} can also be calculated from the plot because the intersection of the graph with the y axis is equal to $1/V_{max}$. The Lineweaver–Burk plot is particularly useful in accurately predicting the K_m and V_{max} of an enzyme. It is also very useful in distinguishing between competitive and non-competitive inhibitors.

Non-competitive inhibitors do not alter K_m but reduce V_{max} and hence the curve is shifted to C (in the diagram opposite). **Reversible competitive inhibitors on the other hand increase K_m but do not alter V_{max}** (the curve will be shifted to D in the diagram).

Answers

41. 1 – J, 2 – K, 3 – G, 4 – H, 5 – D

42. Concerning nerve cells (true or false?)

a. The nerve cell membrane is more permeable to Na^+ than K^+
b. Nerve axons are all myelinated in the peripheral nervous system
c. A decrease in extracellular Ca^{2+} increases the excitability of the cell membrane
d. The distribution of Na^+ channels is uniform across the entire membrane surface
e. Saltatory conduction takes place in unmyelinated nerves

43. With regard to distribution of ions across the membrane of a neuronal axon

a. The concentration of Na^+ is higher intracellularly than extracellularly
b. The concentration of K^+ is higher extracellularly than intracellularly
c. The concentration of Ca^{2+} is higher intracellularly than extracellularly
d. The concentration of Cl^- is higher intracellularly than extracellularly
e. The pH is higher intracellularly than extracellularly

44. From the options listed as A–J, select the most appropriate values to match the normal physiological conditions listed in 1–5

Options

A. 200 mmol/L
B. 10 mmol/L
C. 140 mmol/L
D. 140 μmol.L
E. 1.4 mol/L
F. 3 mmol/L
G. 4.5 μmol/L
H. 4.5 mmol/L
I. 0.6 mmol/L
J. 110 mmol/L

1. Intracellular Na^+
2. Extracellular Ca^{2+}
3. Extracellular Cl^-
4. Extracellular Na^+
5. Extracellular K^+

EXPLANATION: NERVES AND ACTION POTENTIALS

The higher permeability of the neuronal membrane to K^+ ions allows the generation of the resting membrane potential. Not all nerves in the peripheral nervous system are myelinated: **nerve fibres of class A and B are myelinated, C fibres are not**. A lower extracellular Ca^{2+} concentration decreases the amount of polarization necessary to generate an action potential, and hence destabilizes the membrane. The distribution of Na^+ channels is variable – 2000 to 12 000 per mm^2 at nodes of Ranvier in myelinated nerves, but only 20 to 75 per mm^2 at axon terminals. **Saltatory conduction** relies on insulating myelin sheathes between nodes of Ranvier to allow the depolarization to 'jump' quickly from node to node.

Although subject to variation in living tissue, the commonly quoted average values for intra- and extracellular ion concentrations are as follows:

	Extracellular	**Intracellular**
$[Na^+]$ (mmol/L)	140	10
$[K^+]$ (mmol/L)	4.5	140
$[Ca^{2+}]$ (mmol/L)*	3	1
$[Cl^-]$ (mmol/L)	110	3
pH	7.35	7

*Total including bound Ca^{2+}

Answers

42. F F T F F
43. F F F F F
44. 1 – B, 2 – F, 3 – J, 4 – C, 5 – H

45. Action potentials in nerves

- **a.** During the action potential, Na^+ conductance of the membrane increases
- **b.** The membrane potential remains negative throughout for the duration of the action potential
- **c.** Opening of voltage-gated K^+ channels prolongs depolarization
- **d.** A hyperpolarized neuron membrane has a positive membrane potential
- **e.** During the action potential, Na^+ enters the cell and K^+ leaves the cell

46. Regarding action potentials. Select the most appropriate labels for the points labelled 1–5 on the diagram of an action potential in a neuron axon

Options

- **A.** Voltage-gated K^+ channels close
- **B.** Voltage-gated Na^+ channels open
- **C.** Ca^{2+}-dependent Ca^{2+} channels open
- **D.** Electrical gradient for Na^+ changes direction
- **E.** Depolarization
- **F.** Hyperpolarization
- **G.** –30 mV
- **H.** 30 mV
- **I.** 60 mV
- **J.** –70 mV

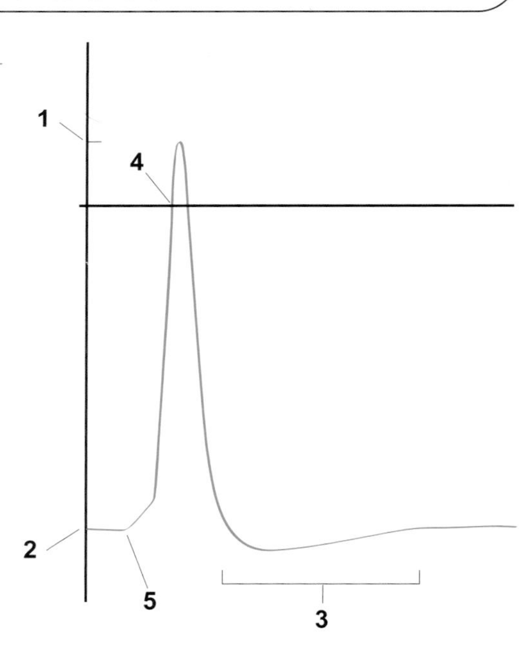

47. Consider the following questions

- **a.** Local anaesthetic drugs block voltage-gated Na^+ channels. How would this affect action potentials in nerves?
- **b.** Local anaesthetic drugs can only block Na^+ channels from inside the cell. What characteristics of a drug molecule would allow it to cross the membrane easily?

EXPLANATION: ACTION POTENTIALS

An **action potential** begins when the membrane is depolarized by about 15 mV. The resting membrane potential of a neuron axon is about −70 mV; a depolarization to about −55 mV will initiate an action potential. At this point, voltage-gated Na^+ channels open, the membrane conductance to Na^+ increases and the membrane depolarizes rapidly. The equilibrium potential for Na^+ is about +60 mV, but this level is not reached, mainly because the Na^+ channels quickly switch to an inactive state, but also because as the membrane potential becomes positive, the direction of the electrical gradient switches, impairing the inflow of Na^+. An action potential typically peaks at approximately 30 mV. Voltage-gated K^+ channels then open, and the increased K^+ conductance of the membrane acts to restore the polarization of the membrane, though there is some overshoot and the membrane becomes slightly hyperpolarized (with a more negative potential than at rest) for a short time. During an action potential then, membrane conductance of both Na^+ and K^+ increases.

If Na^+ channels were blocked by an anaesthetic, no action potential would be possible (47a). Such blocking of Na^+ channels prevents conduction in sensory nerves and hence blocks pain signals. Molecules that cross membranes easily are small and uncharged. Larger, charged molecules would require transport across the enzyme either by facilitated diffusion or active transport, therefore local anaesthetic drugs should ideally be small uncharged molecules (47b).

Answers

45. T F F F T
46. 1 – H, 2 – J, 3 – F, 4 – D, 5 – B
47. See explanation

48. Myelination of nerves

a. Nodes of Ranvier are approximately 1 μm apart
b. Myelination increases the rate of action potential propagation
c. Only axons within the peripheral nervous system are myelinated
d. In the central nervous system, myelin sheaths are formed by astrocytes
e. In the peripheral nervous system, myelin sheaths are formed by Schwann cells

49. Select the most appropriate answer to questions 1–5 from the options below

Options

A. 0.01
B. 1
C. 0.1
D. 0.2
E. 2
F. 200
G. 12
H. 120
I. 20
J. 1000

1. What is the typical conduction velocity in unmyelinated C fibres (m/s)?
2. What is the approximate fastest conduction rate found in human nerve fibres (m/s)?
3. What is the typical duration of muscle depolarization in cardiac muscle (ms)?
4. What is the typical duration of an action potential in a neuron (ms)?
5. What is the largest diameter of neuron axons found in humans (μm)?

50. In striated muscle

a. The resting membrane potential is approximately –90 mV
b. The action potential lasts approximately 200 ms
c. Membrane depolarization results from Na^+ influx
d. Depolarization of sarcoplasmic reticulum membrane leads to T-tubule Ca^{2+} release
e. The action potential propagates by saltatory conduction

51. Briefly describe the events that produce an action potential in cardiac muscle

CNS, central nervous system; PNS, peripheral nervous system

EXPLANATION: NERVES AND MUSCLES

Nodes of Ranvier are breaks in the continuity of nerve myelination that occur approximately 1 mm apart and that allow saltatory conduction to occur. **Saltatory conduction** cannot occur in non-myelinated nerve fibres. Myelination and the saltatory conduction it creates greatly increase the velocity of action potential propagation. Nerves within the central and peripheral nervous systems may be myelinated; in the CNS myelin sheaths are formed by oligodendrocytes and in the PNS by Schwann cells.

Conduction rates in human neuronal axons range between 0.5 and 120 m/s. **A alpha fibres** are myelinated and carry proprioception and somatic motor signals; they are between 12 and 20 μm in diameter. A alpha fibres are the fastest conductors in the human body with maximum rates of conduction of about 120 m/s.

The resting membrane potential of striated muscle is about −90 mV and the action potential is 2–4 ms long. As in nerve fibres, it is the opening of Na^+ channels and influx of Na^+ ions that depolarizes the membrane. The action potential travels into T-tubules, which stimulates release of Ca^{2+} from the terminal cisterns of the sarcoplasmic reticulum.

As in nerve and striated muscle cells, the initial depolarization of cardiac muscle is due to the opening of Na^+ channels. The resting membrane potential of cardiac muscle cells is about −90 mV. The depolarization is short, about 2 ms, but the plateau phase of the cardiac muscle action potential lasts about 200 ms or more. **The plateau phase is due to opening of voltage-gated Ca^{2+} channels**. The resting membrane potential is restored as the Ca^{2+} channels close, and through the action of a number of K^+ channels (51).

Answers

48. F T F F T
49. 1 – B, 2 – H, 3 – F, 4 – E, 5 – I
50. T F T F F
51. See explanation

SECTION 2

CELL BIOLOGY

- CELLS 38
- EUKARYOTES AND PROKARYOTES 38
- THE GOLGI APPARATUS 40
- ENDOPLASMIC RETICULUM 42
- MITOCHONDRIA 44
- CILIA 44
- LYSOSOMES 46
- ORGANELLES 48
- INTRACELLULAR STRUCTURES 48
- PARTS OF A CELL 50
- ENDOCYTOSIS AND EXOCYTOSIS 50
- CYTOSKELETON (i) 52
- CYTOSKELETON (ii) 54
- CELL JUNCTIONS 56
- THE CELL CYCLE (i) 58
- THE CELL CYCLE (ii) 60
- CELL DIVISION 62
- HISTOLOGY 62
- EPITHELIA 64
- HISTOLOGY – MUSCLE 66
- CELLS OF THE BLOOD 68
- CELLULAR BASIS OF PATHOLOGY 68

SECTION 2 CELL BIOLOGY

1. Concerning cells (true or false?)

a. Some cells in the human body have more than one nucleus
b. Some cells in the human body have no nucleus
c. Organelles cannot be visualized with the light microscope
d. All DNA in human cells is stored in the nucleus
e. All cells in the human body reproduce approximately every month

2. List three differences between eukaryotic cells and prokaryotic cells

3. Concerning prokaryotes

a. Metabolism may be either anaerobic or aerobic
b. The genomic DNA is contained within the nucleus by a double membrane
c. Messenger RNA is synthesized in the cytoplasm
d. Mitochondria are absent
e. A cytoskeleton is absent

4. The following are eukaryotic organisms

a. Cats **b.** Trees **c.** Baking yeast **d.** Blue-green algae **e.** Bacteria

5. True or false? Eukaryotes

a. May be surrounded by a cell wall
b. Have lysosomes that are derived from the Golgi apparatus
c. In general replicate more slowly than prokaryotes
d. Cannot survive for a significant time without a nucleus
e. Rely on mitochondria for all their energy requirements

6. Five of the following structures may occur in prokaryotes. Identify which ones

Options

A. Flagella
B. Nucleus
C. Cell wall
D. Glycocalyx
E. Cytoplasm
F. Endoplasmic reticulum
G. DNA
H. Mitotic spindle
I. Golgi apparatus
J. Mitochondria

ATP, adenosine triphosphate

EXPLANATION: CELLS

Cells in the human body generally have a single nucleus. However, skeletal muscle cells form from many individual cells which join before birth and the resulting combined cells have many nuclei, while mature erythrocytes have no nuclei. The nucleus of a cell is usually clearly visible with the light microscope. The nucleus is the primary store of DNA, but mitochondria contain their own DNA. The rate at which cells divide varies; many cells reproduce very slowly – hepatocytes every year or so, endothelial cells lining the cornea and skeletal muscle cells not at all. **Prokaryotes** differ from **eukaryotes** in the following ways (2) (this list is not exhaustive):

	Prokaryotes	Eukaryotes
Nucleus	No	Yes
Organelles	None	Yes
Transcription occurs in	Cytoplasm	Nucleus
Translation occurs in	Cytoplasm	Cytoplasm
Cytoskeleton	No	Yes
Genome	Single circular molecule of DNA	Organized into linear chromosomes
Metabolism	Anaerobic or aerobic	Aerobic
Cell size	Generally less than 10 μm	Generally greater than 10 μm

All animals, plants and fungi are eukaryotes, and all bacteria are prokaryotes. Although complex multicellular organisms are eukaryotic, some eukaryotic organisms exist as single-celled organisms, including yeasts and amoebae.

EXPLANATION: EUKARYOTES AND PROKARYOTES

Like bacteria, some **eukaryotic** cells contain cell walls (e.g. plant cells). The cell walls of plants however are made of cellulose, while those of **prokaryotes** are made of a distinct polysaccharide wall. Eukaryotes, unlike prokaryotes, contain several membrane-bound **organelles**, e.g. **nucleus**, the **Golgi apparatus** and **lysosomes**. Lysosomes, which are derived from the Golgi, contain hydrolytic enzymes which digest endocytosed material and cellular waste products. Whilst mitochondria provide a vast amount of energy in the form of ATP by **oxidative phosphorylation**, some energy is also derived from **glycolysis** occurring in the cytoplasm. This allows cells to survive for a period of time under hypoxic conditions.

Although the nucleus is essential for cellular replication and the generation of new protein components, it is not necessary for cell survival; both the platelet and erythrocyte survive and function for long periods in the absence of a nucleus – the erythrocyte for periods of up to 120 days. **Bacteria have no organelles**. Following DNA replication in bacterial cells, each strand of circular DNA remains attached to the plasma membrane and separates as the cell grows before dividing by simple fission.

Answers

1. T T F F F
2. See explanation
3. T F T T T
4. T T T F F
5. T T T F F
6. 1 A 2 C 3 D 4 E 5 C

7. The Golgi apparatus (true or false?)

a. The *cis* face of the Golgi apparatus is adjacent to the rough endoplasmic reticulum
b. The Golgi apparatus, *trans* face is aligned toward the cell membrane
c. All cisternae of the Golgi apparatus are interconnected and share a common lumen
d. The number of Golgi stacks per cell is similar between cell types
e. The enzymes found in the Golgi apparatus differ between cisternae

8. Theme – The Golgi apparatus. Complete the following passage using terms from the list of options below

Options

A. Regulated exocytosis
B. *Cis*
C. Regulated endocytosis
D. Pancreatic beta-cells
E. Constitutive exocytosis
F. Vesicles
G. Endosomes
H. Cisternae
I. *Trans*
J. Smooth muscle cells
K. Intron
L. Exon
M. Oocytes
N. Adipocytes

The Golgi apparatus is an organelle composed of many disc-shaped membrane-bound compartments called 1 arranged in stacks. Each stack has a 2 face which serves as the entry point to the Golgi, and a 3 face which serves as the exit for modified proteins and is oriented to face the plasma membrane. Specialized secretory cells such as 4 contain large amounts of Golgi apparatus. Concentrated secretory proteins are stored in vesicles before being released in response to external stimuli by the process of 5.

EXPLANATION: THE GOLGI APPARATUS

The **Golgi apparatus** is a membrane-bounded organelle which is responsible for modification of lipids and proteins. The Golgi apparatus consists of many distinct **cisternae** – they have separate lumens and contain different proteins depending on their position within the Golgi. The cisternae are arranged in stacks, and cells have varying numbers of stacks depending on their function.

The Golgi apparatus has a very important polarity: the *trans* face is aligned toward the plasma membrane and the *cis* face is adjacent to the rough endoplasmic reticulum.

Answers

7. T T F F T
8. 1 – H, 2 – B, 3 – I, 4 – D, 5 – A

9. Regarding the rough endoplasmic reticulum, which of the following statements are true?

- **a.** Disulphide bonds between cysteine residues of a protein form in the endoplasmic reticulum
- **b.** Disulphide bonds do not form in the cytosol
- **c.** Almost all glycosylation of proteins takes place in the endoplasmic reticulum
- **d.** Glycosylation of proteins in the endoplasmic reticulum always begins with attachment of a 14-unit oligosaccharide
- **e.** *N*-linked oligosaccharides are the most commonly occurring form of protein glycosylation

10. With regard to the endoplasmic reticulum

- **a.** Some proteins are transported into the endoplasmic reticulum while they are still being synthesized
- **b.** Proteins are directed into the endoplasmic reticulum by a signal sequence of hydrophilic amino acids
- **c.** Transmembrane proteins must be fully transported into the endoplasmic reticulum lumen before relocation to the membrane
- **d.** Stop sequences recognized by endoplasmic reticulum channels are usually hydrophobic
- **e.** Once located in the endoplasmic reticulum membrane, transmembrane proteins cannot alter their orientation

11. Sorting of proteins

- **a.** Proteins are usually sorted by recognition of a signal sequence formed by an attached carbohydrate unit
- **b.** Proteins may be unfolded during their passage into some organelles
- **c.** Movement of proteins from the cytosol to the endoplasmic reticulum requires specific transport proteins
- **d.** Signal sequences may be removed after proteins reach their target organelle
- **e.** Proteins moving from the cytosol to the nucleus must unfold during transport

ER, endoplasmic reticulum; RER, rough endoplasmic reticulum; SER, smooth endoplasmic reticulum

EXPLANATION: ENDOPLASMIC RETICULUM

Hepatocytes have large amounts of **smooth endoplasmic reticulum** (SER), which is involved in modification and detoxification of hydrophobic compounds and rough endoplasmic reticulum (RER) which is involved in synthesis of proteins destined to be targeted to organelles or to function outside the cell. The disulphide bonds that stabilize protein structure form in the RER – the oxidation of pairs of cysteine side chains cannot take place in the cytosol. Almost all glycosylated proteins have their oligosaccharides attached in the ER. The process of glycosylation in the ER always begins with a 14-unit oligosaccharide being transferred to the amino group of an asparagine's side chain. Such *N*-linked oligosaccharides are the most commonly occurring glycosylations.

Membrane-bound **ribosomes** on RER synthesize proteins and feed them into the ER lumen while they are being formed. Proteins are directed to the ER by a sequence of **hydrophobic** amino acid residues. Transmembrane proteins remain in the membrane throughout their synthesis by ER ribosomes. A sequence of hydrophobic amino acid residues is recognized as a stop sequence by the ER translocation channel, which opens a 'side door' that releases the protein sideways into the membrane. The orientation of membrane proteins formed in this way is fixed.

Proteins are sorted by recognition of an amino acid sequence. Proteins are unfolded as they pass into certain organelles, for example **mitochondria** and **peroxisomes**. In order to enter the ER from the cytosol, proteins are unfolded and passed into the ER lumen by specialized translocator membrane proteins. The signal sequence that initiates transport into an organelle may be removed once the protein has been transported. **Proteins are able to pass into the nucleus in their folded state**.

Answers

9. T T T T T
10. T F F T T
11. F T T T F

12. Theme – Mitochondria. Fill in the gaps in the passage below with selections from the list A–N.

Options

A. Male; **B.** Youngest; **C.** Vesicles; **D.** Microvilli; **E.** Cristae; **F.** Cytosol; **G.** Equal to; **H.** Glycogenolysis; **I.** Oxidative phosphorylation; **J.** Glycolysis; **K.** Matrix; **L.** Lower than; **M.** Higher than; **N.** Female

Mitochondria are organelles bounded by two membranes. The inner membrane has many projections forming folds called 1 that increase the surface area of the membrane. The space bounded by the inner membrane is called the 2. Mitochondria produce ATP by the process of 3. The pH at the external face of the inner membrane is 4 the pH at the internal face. Mitochondrial DNA is only inherited from the 5 parent.

13. Concerning mitochondria

a. All mitochondrial membrane proteins are synthesized from mitochondrial DNA
b. Proteins can enter the mitochondrial matrix from the cytosol without unfolding
c. Proteins are transported simultaneously across the inner and outer mitochondrial membranes at points of contact
d. Most mitochondrial membrane phospholipids are synthesized in the endoplasmic reticulum
e. The high pH of the mitochondrial matrix prevents protein folding

14. Regarding cilia (true or false?)

a. They are part of the extracellular matrix
b. They can propel fluid over an epithelial surface
c. The microtubule core has an outer ring of 11 paired microtubules
d. They do not function normally in cystic fibrosis
e. They do not function normally in Kartagener's syndrome

15. Theme – Cilia. Fill in the gaps in the following passage by selecting from the list A–J

Options

A. Vas deferens; **B.** Fallopian tube; **C.** Myosin; **D.** Dynein; **E.** Intermediate filaments; **F.** Microtubules; **G.** Flagella; **H.** Pseudopods; **I.** Urinary tract; **J.** Respiratory tract

Cilia are hair-like projections found on the surface of many cells. At their core is a bundle of 1 which slide over each other as the cilium beats. The structure of the core is identical in the longer 2 found for example on spermatozoa. The beating is produced by associated motor proteins such as ciliary 3. Ciliated cells are typically found in epithelia of the 4 and 5.

ATP, adenosine triphosphate

EXPLANATION: MITOCHONDRIA

Mitochondria are the site of **oxidative phosphorylation** and produce most of the ATP in a cell. They are bounded by a double membrane, dividing the interior into two spaces – the **intermembranous space** and the **matrix**. Mitochondria are cylindrical and vary in size between 0.5 and 2 μm. Mitochondria carry their own DNA, all of which is inherited from the maternal line. Most mitochondrial proteins however derive from nuclear DNA. Proteins unfold in order to enter mitochondria and are fed through both inner and outer membranes simultaneously by translocator proteins. Mitochondrial membrane phospholipids are transported from the endoplasmic reticulum. The mitochondrial matrix has a relatively neutral pH; proteins refold within the matrix after transport across the membrane.

EXPLANATION: CILIA

Cilia are hair-like projections from a cell which are about 0.25 μm in diameter. Their beating can move fluid over **epithelia**; for example, mucus is transported upwards from the respiratory tract. The microtubule core consists of a **9 + 2 arrangement** – nine microtubule pairs surrounding two single microtubules. Cilia function normally in cystic fibrosis, but respiratory problems develop due to abnormalities in the consistency of mucus. **Kartagener's syndrome** is caused by the absence of **axonemal dynein**, which is responsible for the beating of cilia. The loss of motile cilia leads to respiratory disease as mucus is not cleared from the respiratory tract. It also leads to infertility due to failure to produce motile sperm.

Immotile cilia can also lead to **situs inversus** – the condition in which normal right–left asymmetry of organs such as the heart and liver is lost. This is thought to be largely because of the role of cilia in producing the asymmetry as the organs develop in the embryo.

Answers

12. 1 – E, 2 – K, 3 – I, 4 – L, 5 – N
13. F F T T F
14. F T F F T
15. 1 – F, 2 – G, 3 – D, 4 – J, 5 – B

16. Which of the following are lysosomal enzymes?

Options

- **A.** Acid DNase
- **B.** Catalase
- **C.** Beta-glucuronidase
- **D.** DNA polymerase
- **E.** Beta-galactosidase
- **F.** Reverse transcriptase
- **G.** Alpha-glucosidase
- **H.** Citrate synthase
- **I.** Pyruvate kinase
- **J.** Acid RNase

17. Lysosomal storage disorders (true or false?)

- **a.** Are common, affecting approximately 1 in 200 live births in the UK
- **b.** Are usually fatal before adulthood
- **c.** Result from autosomal dominant inheritance
- **d.** Cause affected individuals to have cells with small, shrunken lysosomes
- **e.** Include diabetes mellitus

18. Tay–Sachs disease

- **a.** Is a lysosomal storage disorder
- **b.** Usually causes death in the late teens
- **c.** Inheritance is X-linked recessive
- **d.** Can be treated with regular growth hormone injections
- **e.** Results from a mutation in an ion channel gene

19. Examples of diseases which derive from abnormal lysosomal function include

- **a.** Tay–Sachs disease
- **b.** Gaucher's disease
- **c.** Adrenoleukodystrophy
- **d.** The Ehlers–Danlos syndromes
- **e.** Pemphigus

EXPLANATION: LYSOSOMES

Lysosomes contain hydrolytic enzymes which break down macromolecules. There are many lysosomal enzymes which target specific macromolecules – **nucleases**, **glycosidases**, **proteases** and **lipases**. Catalase is a peroxisomal enzyme and DNA polymerase is involved in DNA synthesis. Reverse transcriptase is a DNA-synthesizing enzyme found in retroviruses.

Lysosomal storage disorders are rare, affecting approximately 1 in 4800 live births in the UK. They are usually fatal before adulthood, and result from recessive inheritance. The conditions lead to trapping of large molecules within lysosomes which become enlarged. Diabetes mellitus is not a lysosomal storage disorder.

Tay–Sachs disease is one of the lysosomal storage disorders that results from a lack of the alpha-chain of the enzyme hexosaminidase A. The onset of symptoms occurs within the first year of life with death occurring at 3–4 years of age. The inheritance pattern of Tay–Sachs disease is autosomal recessive and there is currently no treatment for it. Gaucher's disease is also a lysosomal storage disorder. Adrenoleukodystrophy arises as a consequence of a faulty peroxisomal enzyme. The various **Ehlers–Danlos** syndromes are due to mutations in the genes coding for collagen. Pemphigus is an autoimmune skin condition in which antibodies to the proteins found in desmosomal junctions are formed, leading to widespread blistering.

Answers

16. 1 – C, 2 – G, 3 – E, 4 – J, 5 – A
17. F T F F F
18. T F F F F
19. T T F F F

20. Concerning organelles

- **a.** Primary lysosomes are formed by budding from smooth endoplasmic reticulum
- **b.** The enzyme catalase is found abundantly in peroxisomes
- **c.** Tay–Sachs disease results from a defect in smooth endoplasmic reticulum phospholipid synthesis
- **d.** Hepatocytes do not have a smooth endoplasmic reticulum
- **e.** The inner mitochonrdrial membrane is more permeable than the outer membrane to large molecules

21. Regarding organelles

- **a.** The nuclear envelope consists of a single membrane layer
- **b.** The pH inside lysosomes is approximately 8
- **c.** Rough endoplasmic reticulum is coated with ribosomes
- **d.** Glycolysis takes place in mitochondria
- **e.** The pH in the matrix of mitochondria is higher than in the intermembrane space

22. Concerning organelles, which of the following statements are true?

- **a.** Lysosomes and peroxisomes contain identical enzymes but differ in their pH
- **b.** Secondary lysosomes have H^+ ATPase pumps in their membranes
- **c.** Acid hydrolases are peroxisomal enzymes
- **d.** Primary lysosomes have H^+ ATPase pumps in their membranes
- **e.** Lysosomes are bounded by a double layer of plasma membrane

23. Theme – organelles. From the options listed as A–H, select the structures that best match statements 1–5

Options

A. Centrosome; **B.** Nucleus; **C.** Nucleolus; **D.** Peroxisome; **E.** Lysosome; **F.** Mitochondrion; **G.** Endoplasmic reticulum; **H.** Golgi apparatus

1. This structure has an internal pH of approximately 5
2. This structure is comprised of stacks of cisternae
3. This structure generates ATP by oxidative phosphorylation
4. Oxidation of toxins takes place in this structure
5. This structure is an organizing centre for microtubules

24. Regarding cytoplasmic inclusions

- **a.** Lipofuscin is found only in the cells of neonates
- **b.** Lipofuscin is not found in nerve or liver cells
- **c.** Lipid may be stored in non-membrane-bound vacuoles
- **d.** Glycogen is stored in membrane-bound organelles
- **e.** Accumulation of lipids in liver cells may be caused by drinking alcohol in excess

---

ER, endoplasmic reticulum; ATP, adenosine triphosphate

EXPLANATION: ORGANELLES

The nuclear envelope is a double membrane structure, with the outer membrane continuous with that of the ER. Rough ER has a grainy appearance on electron microscopy because of the high concentration of membrane-bound ribosomes.

Glycolysis takes place in the cytosol. The proton pump action of the electron transport chain proteins maintains a pH gradient across the inner mitochondrial membrane – the pH in the matrix is approximately 8, while in the intermembrane space it is approximately 7.

The degradative enzymes within lysosomes are optimally active in **acidic** conditions in contrast to those found in peroxisomes, so the pH in lysosomes is maintained at approximately 5. Primary lysosomes bud from the Golgi apparatus, but do not have H^+ ATPase pumps in their membrane. Secondary lysosomes are formed by fusion of primary lysosomes with endosomes, which contain specific membrane proteins such as the **H^+ ATPase pump**. Lysosomes are bounded by a single plasma membrane. Tay–Sachs disease results from a defect in a lysosomal enzyme involved in the degradation of glycolipids.

Catalase can account for up to 40 per cent of peroxisomal protein, and catalyses the conversion of H_2O_2 to $O_2 + H_2O$. The outer membrane of mitochondria is far more permeable than the inner membrane, allowing the passage of larger molecules up to 10 000 Da.

EXPLANATION: INTRACELLULAR STRUCTURES

The Golgi apparatus is composed of stacks of disc-shaped structures called cisternae. **Mitochondria** are the site of **oxidative phosphorylation** and the proteins of the electron transport chain are arranged on the inner membrane. Peroxisomes are involved in oxidative processes, particularly the oxidation of long-chain fatty acids and toxins. The **microtubular network** of the cytoskeleton is arranged around and centred on the **centrosome**, which doubles in forming the mitotic spindle.

Lipofuscin is a pigment derived from products of cell degradation that accumulates in old age, it is mostly found in nerve, cardiac muscle and liver cells. Lipids may be stored within cells, particularly adipocytes, as large vacuoles that are not bounded by membrane. Intracellular glycogen is not stored within membrane-bound compartments. The commonest cause of fat accumulation in hepatocytes is a high alcohol intake.

Answers

20. F T F F F
21. F F T F T
22. F T F F F
23. 1 – E, 2 – H, 3 – F, 4 – D, 5 – A
24. F F T F T

25. Identify the components of the cell labelled 1–10 on the diagram opposite

Options

A. Golgi apparatus
B. Mitochondrion
C. Nucleus
D. Cytosol
E. Lysosome
F. Centriole
G. Rough endoplasmic reticulum
H. Nucleolus
I. Peroxisome
J. Microvilli

26. Regarding vesicles

a. They are surrounded by a double layer of plasma membrane
b. They usually develop with a protein coat
c. Clathrin is a protein found on the luminal surface of developing vesicles
d. Dynamin is a protein involved in pinching off vesicles from membranes
e. Vesicular transport occurs both into and out of cells

27. Concerning endocytosis

a. Most cells are capable of phagocytosis
b. Phagocytosis involves uptake of larger particles than pinocytosis
c. In any given time period the volume of endocytosed matter is about twice that of exocytosed matter
d. Pinocytosis uptakes molecules selectively according to needs of the cell
e. Receptor-mediated endocytosis can be an entry point for viruses

28. With regard to exocytosis

a. Constitutive exocytosis supplies lipid and protein to the plasma membrane
b. Regulated exocytosis only occurs in specialized secretory cells
c. Secretory vesicles bud from the *cis* face of the Golgi apparatus
d. Proteins secreted by constitutive exocytosis tend to aggregate before leaving the Golgi
e. Constitutive exocytosis takes place in all eukaryotic cells

GTP, guanosine triphosphate; HIV, human immunodeficiency virus

EXPLANATION: PARTS OF A CELL

It is important to be able to recognize a number of intracellular structures. Most are not visible under light microscopy, but the most commonly encountered structures are those shown in the diagram opposite. Identifying features are described below:

- **Nucleus** – large pigmented structure bounded by a double membrane and containing the darker nucleolus
- **Rough endoplasmic reticulum** – Membrane continuous with that of nucleus. 'Studded' with ribosome granules
- **Mitochondria** – Cylindrical structures with characteristic infolding of inner membrane forming cristae
- **Golgi apparatus** – multiple layers of cisternae with budding and transport vesicles between layers
- **Centriole** – a paired cylinder structure associated with the centrosome and the microtubular network

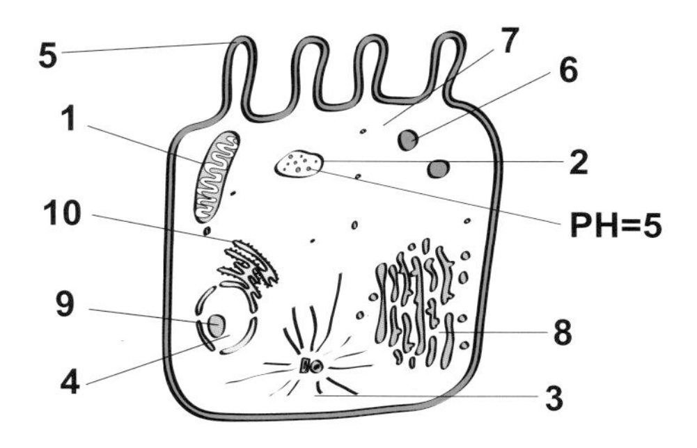

EXPLANATION: ENDOCYTOSIS AND EXOCYTOSIS

Vesicles are surrounded by a single membrane layer. When a vesicle buds from a membrane, it usually has a coat of protein on the cytosolic aspect of the membrane; this coat is lost after budding. A typical protein that forms a vesicular coat during budding is **clathrin**, which is attached to the cytosolic side of the vesicle membrane. Dynamin is a GTP-binding protein which forms around the neck of an almost completely formed vesicle, and separates the vesicle from its parent membrane. Vesicular transport is involved in export and import of cellular material.

Very few cells in multicellular organisms are capable of **phagocytosis**, which is the uptake of **large** particles into a cell. All cells are capable of **pinocytosis**, which involves the uptake of fluid and particles **<150 nm** in diameter, whereas phagocytosis involves ingestion of much larger particles. The amount of fluid taken up by the cell by **endocytosis** must be equal to the amount discharged by **exocytosis**, or the cell would shrink or enlarge. Pinocytosis is a **non-selective** process. Receptor-mediated uptake is exploited by some viruses; HIV enters its target cells in this way. **Exocytosis** occurs in all eukaryotic cells and is the pathway that supplies lipid and protein to the plasma membrane. **Regulated exocytosis** is the pathway by which specialized secretory cells release their products. In order to concentrate secretions, the proteins secreted by regulated exocytosis aggregate in the *trans*-Golgi network. The *trans* side of the Golgi apparatus faces the cell membrane, and it is from this side that secretory vesicles bud.

Answers

25. 1 – B, 2 – E, 3 – F, 4 – C, 5 – J, 6 – I, 7 – D, 8 – A, 9 – H, 10 – G
26. F T F T T
27. F T F F T
28. T T F F T

29. True or false? The cytoskeleton plays an important role in

a. Cell movement
b. Cell division
c. Determining cell shape
d. Gene expression
e. Axonal transport

30. Theme – the cytoskeleton. Match the components of the cytoskeleton to the descriptions below

Options

A. Microtubules
B. Microvilli
C. Actin filaments
D. Myosin
E. Intermediate filaments
F. Spectrin
G. Centrosome
H. Kinetochore
I. Laminin
J. Flagella

1. These fibres are the main component of the mitotic spindle
2. This structure is an organizing centre for microtubules
3. Networks of these structures attach to desmosomes
4. Interaction of actin filaments with this protein causes muscle contraction
5. These structures have a core of microtubules arranged in a 9 + 2 formation

31. Concerning the cytoskeleton

a. Intermediate filaments are about 10 nm thick
b. Centrosomes are organizing centres for microtubules
c. Intermediate filaments are composed of alpha- and beta-tubulin molecules
d. Microtubules are hollow
e. Actin filaments are involved in muscle contraction

EXPLANATION: CYTOSKELETON (i)

The **cytoskeleton** functions as the 'bones and muscles' of a cell and is composed of actin filaments, intermediate filaments and microtubules. It allows cells to move and adopt different shapes. **Microtubules** form the mitotic spindle, which is important in cell division. The cytoskeleton does not affect gene expression. Transport of materials in neuronal axons takes place along microtubules.

The three main elements of the cytoskeleton are **actin filaments**, **intermediate filaments** and **microtubules**. Microtubules form the mitotic spindle, and their arrangement centres on the centrosomes. Intermediate filaments attach to desmosomes and hemidesmosomes, whereas actin filaments attach to adherens junctions. The two proteins involved in contraction of muscle are actin and myosin; as myosin 'walks' along actin, muscle fibres contract. Flagella and cilia have a core of microtubules arranged as a ring of pairs surrounding two central fibres.

Actin filaments are approximately **7 nm** thick, **intermediate filaments 10 nm** thick and **microtubules** about **25 nm** thick. Centrosomes organize groups of microtubules that grow out from them into the cytoplasm. Microtubules are hollow and composed of alpha- and beta-tubulin subunits. Actin filaments react with myosin filaments in muscle cells to bring about contraction.

Answers

29. T T T F T
30. 1 – A, 2 – G, 3 – E, 4 – D, 5 – J
31. T T F T T

32. Regarding the cytoskeleton

a. Name three main protein filaments that form the cytoskeleton
b. Which of these filament types forms the core of cilia and flagella? What is their arrangement?
c. Name one other function of these filaments
d. Name one function of cilia and one function of flagella in the human body
e. Name one of the consequences of immotile cilia
f. What is the cause of immotile cilia?

33. Concerning myofibrils

a. Thick filaments are formed mainly from myosin
b. Thin filaments are formed mainly from actin
c. The H zone is formed by thick filaments alone
d. The Z band lies in the centre of the I band
e. Actinin holds actin filaments to the Z disc

34. Concerning actin and myosin interaction

a. Muscle contraction is triggered by a sudden fall in cytoplasmic Ca^{2+} concentration
b. Binding of ATP with myosin detaches the myosin head from the actin filament
c. At rest, tropomyosin prevents binding of myosin and actin
d. Tropomyosin undergoes a conformational change when bound to Ca^{2+}
e. Muscle contraction requires hydrolysis of ATP

ATP, adenosine triphosphate

EXPLANATION: CYTOSKELETON (ii)

The main protein filaments that form the cytoskeleton are **actin filaments**, **intermediate filaments** and **microtubules** (33a). **Microtubules** form the core of cilia and flagella; in these structures they are arranged in a **9 + 2 pattern**, with nine double tubules surrounding a central pair (33b). Microtubules also form the mitotic spindle and tracks along which organelles can be ferried by motor proteins, e.g. in **axoplasmic flow** (33c). Cilia move fluid over epithelial surfaces, for example in the respiratory tract. The flagella of spermatozoa are used as a means of propulsion (33d). Immotile cilia cause male sterility (flagella on spermatozoa do not function) and respiratory disease (respiratory tract cilia do not clear mucus from the airways). Immotile cilia can also cause **Kartagener's syndrome** – a triad of sinusitis, bronchitis and situs inversus. Cilia play an important embryological role in forming the left/right asymmetry of the viscera (33e). Immotile cilia are caused by a fault in the motor protein dynein which produces the rhythmic movement of cilia and flagella. There is no defect in the microtubule structure (33f).

Myofibrils are the contractile elements of skeletal muscle cells. The thick filaments are formed by **myosin** and the thin filaments by **actin**. The H zone lies in the centre of the A band and represents the area of thick filaments which do not overlap with thin filaments. The Z band divides **sarcomeres** and is the point of attachment for all the thin filaments, it lies in the centre of the I band, which represents the area of thin filaments that do not overlap with thick filaments. Actinin attaches actin filaments to the Z disc.

Muscle contraction begins with an **increase** in cytoplasmic Ca^{2+} concentration. Binding of ATP with myosin causes the myosin head to detach from the actin filament. Tropomyosin blocks the binding site of myosin and actin when the muscle is at rest. Troponin undergoes a conformational change when bound to Ca^{2+}. Muscle contraction requires a supply of ATP – one molecule of ATP is hydrolysed for each 'step' a myosin head takes along an actin filament.

Answers

32. See explanation
33. T T T T T
34. F T T F T

35. With regard to intercellular junctions

- **a.** Hemidesmosomes connect the intermediate filament networks of adjoining cells
- **b.** Adherens junctions anchor epithelial cells to the basal lamina
- **c.** Binding of cadherins requires the presence of Ca^{2+} in the extracellular fluid
- **d.** Tight junctions prevent leakage of molecules across an epithelium
- **e.** Gap junctions allow the passage of small, water soluble ions from the cytosol to the extracellular space

36. Identify the junction types and associated structures in the diagram opposite

Options

- **A.** Actin bundle
- **B.** Adherens junction
- **C.** Desmosome
- **D.** Microtubules
- **E.** Tight junction
- **F.** Gap junction
- **G.** Intermediate filaments
- **H.** Hemidesmosome

37. Regarding intercellular junctions

- **a.** Occluding junctions prevent lateral migration of membrane-bound proteins
- **b.** Desmosomes connect microtubular networks of neighbouring cells
- **c.** Gap junctions are found at the intercalated discs of cardiac muscle cells
- **d.** Hemidesmosomes join actin bundles of one cell to those of a neighbouring cell
- **e.** Desmosomes allow the passage of certain molecules between neighbouring cells

38. Intercellular junctions

- **a.** Name three types of intercellular junction that may be found between neighbouring epithelial cells
- **b.** For each of the junctions identified in part (a), briefly explain the function of the junction

39. The epithelial basement membrane

- **a.** Is composed of a single layer of cells
- **b.** Contains mostly type I collagen
- **c.** Is impermeable to urea in the kidney glomerulus
- **d.** Contains laminin proteins
- **e.** Is identical in all epithelia

EXPLANATION: CELL JUNCTIONS

There are four classes of junction that form between epithelial cells: **tight junctions**, **adherens junctions**, **desmosomes** and **gap junctions** (39a).

- **Tight junctions** (or occluding junctions) form a seal between cells and occur in bands around cells. These bands form barriers to certain molecules and restrict movement of membrane proteins to fixed areas of the cell membrane.
- **Adherens junctions** tether actin bundles of neighbouring cells. The contact between the cells is through transmembrane proteins, the cadherins which require the presence of Ca^{2+} to bind with each other – hence their name.
- **Desmosomes** join the intermediate filament networks of neighbouring cells. They confer mechanical stability to the epithelial layer by interlinking the cytoskeletons of the cells they join. They do not allow the passage of molecules between cells. Hemidesmosomes anchor intermediate filaments within a cell to the basal lamina.
- **Gap junctions** allow the passage of small molecules such as ions from the cytosol of one cell to the cytosol of its neighbour (39b).

The epithelial basement membrane, or basal lamina, consists of thin sheets of extracellular matrix composed mainly of type IV collagen, laminin, heparan sulphate, entactin and fibronectin. Basement membranes vary in their permeability; in the kidney glomeruli the pores must be large enough to filter waste molecules such as urea.

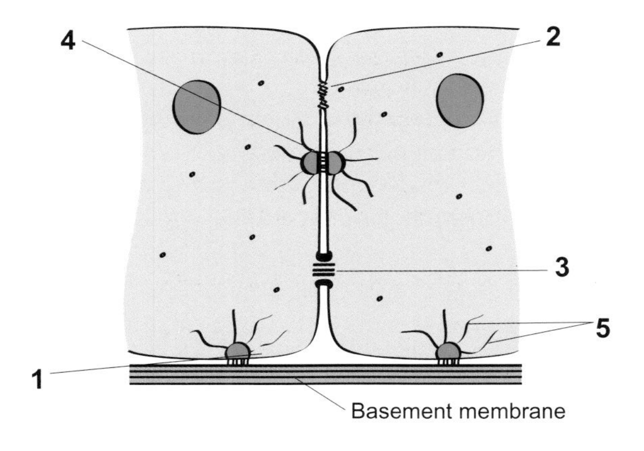

Answers

35. F F T T F
36. 1 – H, 2 – E, 3 – F, 4 – C, 5 – G
37. T F T F F
38. See explanation
39. F F F T F

40. Theme – the cell cycle

a. List, in sequence, the stages of the cell cycle. During which stage is DNA replicated?
b. List, in sequence, the stages of mitosis. During which stage of mitosis does the nuclear envelope dissolve?

41. In mitosis (true or false?)

a. The G_1, S and G_2 phases of the cell cycle occur in interphase
b. DNA is replicated during the prophase
c. Sister chromatids line up at the equator at the start of anaphase
d. The nuclear envelope reforms during telophase
e. The daughter cells are haploid

42. Meiosis

a. Results in daughter cells that are identical to each other
b. Results in cells with a haploid karyotype
c. Describes the process by which gut epithelial cells replicate
d. Results in two daughter cells
e. During oogenesis results in four viable oocytes

43. Regarding prophase

a. Prophase is the first phase of mitosis
b. During prophase the centrosomes move to opposite poles of the cell
c. During prophase the mitotic spindle is formed
d. During prophase the nuclear envelope breaks up into small vesicles
e. During prophase the centrosome is duplicated

44. Concerning prometaphase

a. Prometaphase ends with the breakdown of the nuclear envelope
b. During prometaphase spindle microtubules bind to the chromosomes
c. Kinetochores prevent binding of spindle microtubules to chromosomes
d. During prometaphase the connections between sister chromatids are cut
e. In human cells each kinetochore binds one microtubule

EXPLANATION: THE CELL CYCLE (i)

The stages of the cell cycle are: G_1, S, G_2 and M. DNA replication takes place in the S phase (41a). The stages of mitosis are: prophase, prometaphase, metaphase, anaphase, telophase and cytokinesis. The nuclear envelope dissolves at the start of prometaphase (41b).

Mitosis is the process by which somatic cells replicate. The resultant daughter cells are identical and are **diploid**. The G_1, S and G_2 phases are collectively known as **interphase** (DNA is replicated during the S phase). The nuclear membrane breaks down during prometaphase, **chromatids** align on the equator in metaphase and separate during anaphase. The nuclear membrane reforms after complete separation of the chromatids during telophase and the cell itself divides (cytokinesis).

Meiosis is the process by which diploid germ cells give rise to **haploid** gametes (somatic cells such as gut epithelial cells undergo mitosis). During this process, homologous recombination takes place that results in daughter cells that may be distinct from each other. Whilst in spermatogenesis, four haploid sperm cells are produced, in oogenesis, only one viable ovum is produced. The three other cells resulting from meiosis become polar bodies.

Prophase is the first phase of mitosis during which the duplicated **chromosomes** condense and the mitotic spindle forms. The centrosomes are duplicated during interphase, but they move apart forming the spindle during prophase. The membrane remains intact until prometaphase.

Prometaphase is the second stage of the M phase of the cell cycle. It begins with the breakdown of the nuclear envelope into vesicles. Chromosomes then attach to spindle microtubules – the microtubules attach to the chromosome **kinetochores**. In some organisms each kinetochore binds to only one microtubules, however in humans each kinetochore can bind 30 or more microtubules.

Answers

40. See explanation
41. T F F T F
42. F T F F F
43. T T T F F
44. F T F F F

45. With regard to the cell cycle

a. The G_1 phase lasts for approximately 24 hours for all cells
b. Only certain cells enter the G_0 phase
c. During the S phase new DNA is synthesized
d. The prophase is the first event in the M phase
e. Progress of the cell cycle is regulated by cyclins

46. True or false? During the cell cycle

a. Growth factors (e.g. fibroblast growth factor) can allow a cell to progress from G_0 to G_1 phase
b. DNA is replicated in the G_2 phase
c. Checkpoints exist in the S phase to ensure that all DNA has been copied
d. The cell may undergo apotosis in the G_1 phase
e. Cell division occurs in the M phase

47. Regarding control of the cell cycle

a. Throughout the cell cycle the concentrations of the cyclin-dependent protein kinases remain constant
b. Throughout the cell cycle the enzyme activity of the cyclin-dependent protein kinases remains constant
c. Cyclin has intrinsic enzyme activity
d. The activity of M phase-promoting factor rises just before mitosis
e. There are many varieties of cyclin involved in cell cycle control

EXPLANATION: THE CELL CYCLE (ii)

The G_1 phase of the cell cycle varies in length, depending on the type of cell. The G_0 phase is an exit from the cell cycle, and is entered only by cells that do not divide. During the S phase new DNA is synthesized, i.e. the chromosomes are duplicated. The M phase is divided into five subphases: prophase, prometaphase, metaphase, anaphase and telophase. Cyclins, in combination with cyclin-dependent kinases, regulate the progress of the cell cycle.

Cells in the quiescent G_0 phase may be promoted to enter the cell cycle by **growth factors**. The first step in the cell cycle is the G_1 phase, during which the cell determines if it can undergo division. At this phase, the integrity of the DNA is assessed and repair takes place if necessary. If damage to the DNA is irreversible the cell undergoes **apoptosis**. The cell then enters the S phase, during which DNA is replicated. The G_2 phase then ensues, and here further checkpoints exist to ensure that all the DNA has been copied and that the cell is big enough to undergo division. Division occurs in the final M phase of the cell cycle.

The concentration of **cyclin-dependent kinases** does not change during the cell cycle, however their level of activity changes depending on the concentration of cyclin, which increases and decreases throughout the cycle. Cyclin has no intrinsic enzyme activity. The activity of M phase-promoting factor increases just before mitosis. A variety of cyclins and cyclin-dependent kinases are involved in regulating the progression of the cell cycle.

Answers

45. F T T T T
46. T F F T T
47. T F F T T

48. Cell division. Use the following options to answer the questions below

Options

A. 2
B. 3
C. 46
D. 8
E. 4
F. 1
G. 23
H. 21
I. 42
J. 92

1. How many times does DNA replication takes place in the process of meiosis?
2. How many gametes are produced by one cell entering meiosis?
3. How many daughter cells are produced by one cell entering mitosis?
4. How many chromosomes are contained in a human haploid cell?
5. How many chromosomes are contained in a human diploid cell?

49. When comparing mitosis with meiosis

a. Homologous recombination only occurs in meiosis
b. Both require the mitotic spindle
c. Both process results in two daughter cells
d. Meiosis produces four daughter cells identical to the parent cells
e. Both processes occur in germ cells

50. Theme – histology. Identify the cell type from the description below

Options

A. Peripheral nervous system neuron
B. Keratinized stratified squamous epithelium
C. Cardiac muscle cell
D. Simple columnar epithelium
E. Smooth muscle cell
F. Hepatocyte
G. Oocyte
H. Erythrocyte
I. Central nervous system neuron
J. Endothelial cell

1. This is an excitable cell, parts of which may be myelinated by Schwann cells
2. This is an excitable cell, parts of which may be myelinated by oligodendrocytes
3. This is an example of a haploid cell
4. This cell loses its nucleus as it matures
5. Cells of this type form the epidermis

DNA, deoxyribonucleic acid

EXPLANATION: CELL DIVISION

Meiosis consists of one round of DNA replication and two rounds of cell division, producing four **haploid** cells from one **diploid** cell. **Mitosis** consists of one round of DNA replication and one cell division, producing two **diploid** daughter cells. Diploid cells contain two copies of the 22 autosomes and two sex chromosomes – 46 chromosomes in total. Haploid cells contain 23 chromosomes – a single copy of each autosome and one sex chromosome.

Meiosis produces four daughter cells that are all different from each other and the parent cell, whilst **mitosis** produces two daughter cells that are identical to each other and the parent cell. Homologous recombination occurs only in **meiosis**. Germ cells can undergo **mitosis** (for germ cell replication) and **meiosis** (for the generation of gametes).

EXPLANATION: HISTOLOGY

Nerve cells are often **myelinated** – other cells coat them in an insulating sheath. In the peripheral nervous system, Schwann cells provide the myelin sheath whereas in the central nervous system **myelination** is a function of oligodendrocytes. Gametes are examples of **haploid** cells; they have only one set of **chromosomes** and are formed by the process of meiosis. Erythrocytes are the oxygen-carrying cells of the blood. During their development they lose all organelles, **including the nucleus**. The epidermis is a keratinized stratified squamous epithelium that provides a strong, water-resistant coat for the body.

Answers

48. 1 – F, 2 – E, 3 – A, 4 – G, 5 – C
49. T T F F T
50. 1 – A, 2 – I, 3 – G, 4 – H, 5 – B

51. State whether the following pairings of epithelia types and their locations are true or false

- **a.** Simple squamous epithelium/lining of blood vessels
- **b.** Simple cuboidal epithelium/bladder
- **c.** Ciliated simple columnar epithelium/tongue
- **d.** Stratified squamous epithelium/lining of intestine
- **e.** Transitional epithelium/bladder

52. Regarding epithelia

- **a.** Simple columnar epithelium is composed of many layers of cells
- **b.** Simple cuboidal epithelium is composed of a single layer of cells
- **c.** Stratified squamous epithelium is composed of many layers of cells
- **d.** Cells of pseudostratified columnar epithelium are all in contact with the basement membrane
- **e.** Cells of stratified squamous epithelium appear flattened in the superficial layers

53. Select the most appropriate description of the epithelia types depicted in the diagrams below from choices A–H

Options

- **A.** Pseudostratified columnar epithelium
- **B.** Stratified cuboidal epithelium
- **C.** Stratified squamous epithelium
- **D.** Simple columnar epithelium
- **E.** Simple squamous epithelium
- **F.** Simple cuboidal epithelium
- **G.** Ciliated simple columnar epithelium
- **H.** Urothelium

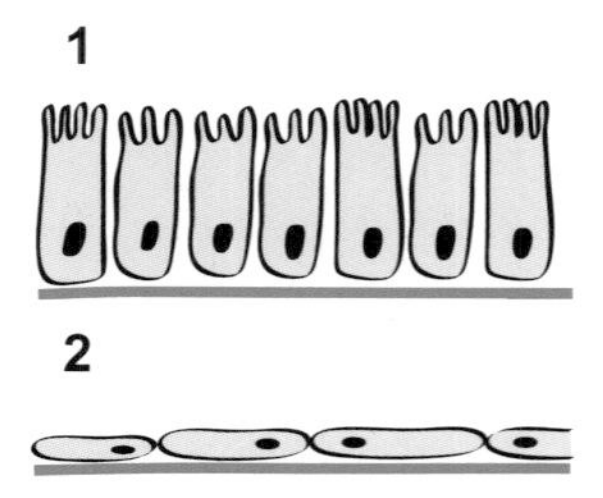

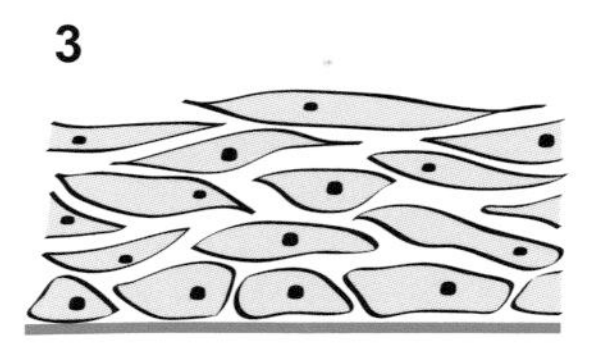

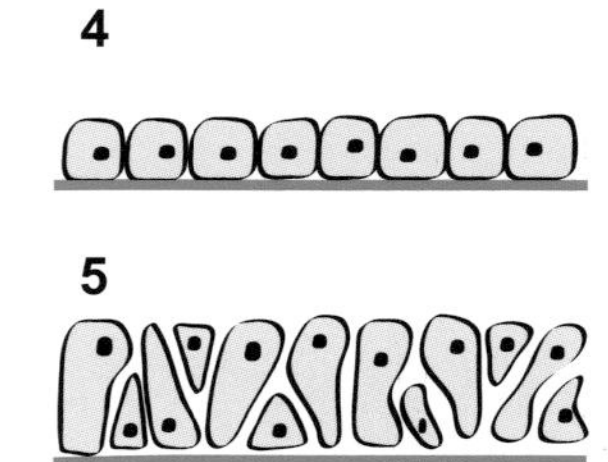

GI, gastrointestinal

EXPLANATION: EPITHELIA

Epithelia are classified according to the shapes and arrangements of their cells. **Stratified epithelia** consist of more than one layer, with cells rising through the layers as the outermost layer is lost. **Simple epithelia** consist of a single layer of cells, all in contact with the basement membrane. **Pseudostratified epithelia** appear to be multilayered, with cell nuclei at different levels, but all cells are in fact in contact with the basement membrane. Epithelial cells may be squamous (flat), cuboidal (roughly square) or columnar (tall and narrow). Additional features of epithelia include cilia which serve to move fluid over the surface of the epithelium.

Simple squamous epithelium is a single flat layer of cells, found lining blood vessels and lymphatics.

Stratified squamous epithelia are composed of many layers of cells which migrate from the basement membrane to the superficial layers, where they appear flattened; they are found in the epidermis of skin.

Simple cuboidal epithelium is a single layer of cube-shaped cells, usually involved in secretion and absorption and found in, for example, kidney tubules and ducts of small glands.

Simple columnar epithelium is composed of a single layer of cells, all of which are attached to the basement membrane. From stomach to anus, the GI tract is lined with non-ciliated simple columnar epithelium.

Ciliated simple columnar epithelium is found in parts of the upper respiratory tract, fallopian tubes and central canal of the spinal cord. The tongue has a stratified squamous epithelium. Transitional epithelium is capable of stretching, and is found lining the bladder and parts of the ureters and urethra.

Answers

51. T F F F T
52. F T T T T
53. 1 – G, 2 – E, 3 – C, 4 – F, 5 – A

54. Theme – histology. Label the diagram of a muscle cell below from the options given

Options

- **A.** M line
- **B.** Sarcoplasmic reticulum
- **C.** Tropomyosin
- **D.** Z disc
- **E.** H zone
- **F.** Intercalated disc
- **G.** Sarcomere
- **H.** I band
- **I.** External lamina
- **J.** A band

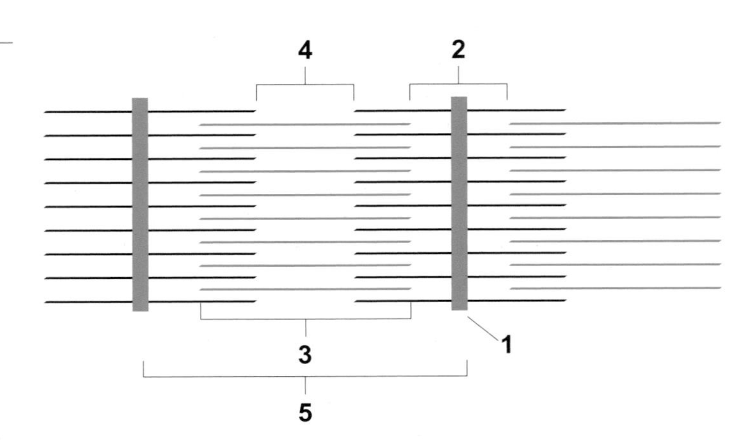

55. True or false? Regarding skeletal muscle

- **a.** Skeletal muscle cells usually have more than one nucleus
- **b.** Each sarcomere is about 2.5 mm long
- **c.** During contraction, neither myosin nor actin filaments change length
- **d.** During contraction the sarcomeres do not change in length
- **e.** The myosin filaments attach to the Z disc

56. Describe how the head of myosin thick filaments moves along actin thin filaments following release of Ca^{2+} in skeletal muscle

ADP, adenosine diphosphate; ATP, adenosine triphosphate

EXPLANATION: HISTOLOGY – MUSCLE

Skeletal muscle has a characteristic banded appearance on light microscopy, each band representing a **sarcomere** within a **myofibril**. They are large cells and have multiple nuclei. On electron microscopy the finer detail of the bands can be seen, the different areas of banding corresponding to the microscopic structure of a sarcomere. Five areas are defined as A bands, H bands, I bands, Z lines and M lines. The A band is the dark area in the middle of the sarcomere formed by the complete length of myosin thick filaments. The H band is a paler region within the A band composed of thick filaments with no overlapping thin filaments. The I band is a pale area formed by thin filaments without overlapping thick filaments. The M band runs down the centre of the H band. The Z line runs down the centre of the I band and represents the Z disc, the point of attachment for the thin filaments.

The contractile elements, **myofibrils**, are composed of repeated arrangements of **sarcomeres**, each approximately 2.5 μm long, though the length depends on the degree of contraction. The myosin and actin filaments that form the thick and thin filaments of the **myofibril** do not change in length during contraction but slide over each other, shortening the sarcomere. Actin filaments are attached to the Z disc that separates sarcomeres.

The process of muscle contraction begins with release of Ca^{2+} from the sarcoplasmic reticulum. Ca^{2+} binds to **troponin**, which changes shape, allowing the myosin head to bind to actin. An ATP molecule binds with the myosin head, causing it to release from the actin filament. The bound ATP induces a conformational change in the myosin head, causing it to assume a 'cocked' position. The ATP is hydrolysed to ADP and inorganic phosphate. The cocked myosin head now forms a bond with the actin filament at a point a little further along the chain from its original position. This new binding releases the inorganic phosphate and initiates the force-generating step as the myosin head returns to its uncocked position, dragging the myosin filament with it. As it moves, the ADP separates from myosin and the cycle begins again from a point approximately 5 nm away from the start of the last cycle.

Answers

54. 1 – D, 2 – H, 3 – J, 4 – E, 5 – G
55. T F T F F
56. See explanation

57. Theme – cells of the blood. Match the blood cell types to the descriptions in list 1–5

Options

A. Erythrocytes **B.** Platelets **C.** Neutrophils **D.** Macrophages **E.** Eosinophils
F. Basophils **G.** Monocytes **H.** Plasma cells **I.** B lymphocytes **J.** T lymphocytes

1. These cells leave the circulation to form mast cells
2. These cells are formed from fragments of megakaryocytes
3. These cells derive from B lymphocytes and actively synthesize immunoglobulin
4. These cells have a characteristic multi-lobed nucleus
5. These cells contain acidophilic granules and have a bilobed nucleus

58. White blood cells

a. All white cells have a nucleus
b. Neutrophils have a multilobed nucleus
c. Neutrophils are capable of phagocytosis
d. Lymphocytes release histamine in response to exposure to allergens
e. Eosinophils contain acidophilic granules

59. Erythrocytes

a. Normal erythrocytes have a lifespan of about 120 days
b. Erythrocytes have a rigid structure and do not deform easily
c. Erythrocytes generate ATP by oxidative phosphorylation
d. Old and defective erythrocytes are disposed of in the spleen
e. Erythrocytes have no nucleus

60. Theme – cell pathology. Match the disorder to the descriptions below

Options

A. Eczema **B.** Familial hyperlipidaemia **C.** Tay–Sachs disease
D. Turner's syndrome **E.** Down's syndrome **F.** Pemphigus
G. Cerebral palsy **H.** Multiple sclerosis **I.** Kartagener's disease
J. Cystic fibrosis

1. This condition results from autoantibodies to desmosome proteins in the skin
2. Features of this disease include situs inversus and immotile cilia
3. This condition often results from abnormal meiosis in the formation of gametes, when a second copy of chromosome 21 is included in either the sperm or egg cell
4. This disease results from autoimmune destruction of myelin sheaths in the central nervous system
5. This disease results from mutation in a gene coding for a lysosomal enzyme

ATP, adenosine triphosphate

EXPLANATION: CELLS OF THE BLOOD

Erythrocytes are the oxygen-carrying cells of the blood. Platelets are derived from megakaryocytes and they play an essential role in haemostasis. There are five types of white blood cell: **neutrophils**, **eosinophils**, **basophils**, **lymphocytes** and **monocytes**. The granulocytes are the neutrophils, eosinophils and basophils.

White blood cells are all nucleated, and neutrophils have a characteristic multilobed nucleus. Neutrophils are involved in the early stages of an inflammatory response and phagocytose bacteria and dead cells. The granules of **basophils** and **mast cells** contain **histamine**, which is released in response to allergen exposure. Eosinophils contain acidophilic granules, basophils contain basophilic granules. On leaving the circulation and entering tissues, basophils become mast cells and monocytes become macrophages. Lymphocytes are involved in generating specific immune responses. B lymphocytes differentiate to form plasma cells that secrete immunoglobulin.

Erythrocytes normally live for about 120 days in the circulation before removal by the spleen. They are flexible and deform to pass through capillaries. Mature erythrocytes have no organelles and are hence unable to produce ATP by oxidative phosphorylation, which normally takes place in mitochondria.

EXPLANATION: CELLULAR BASIS OF PATHOLOGY

Pemphigus is an autoimmune condition resulting from antibodies directed at desmosomes in the skin. It is a serious dermatological condition which causes widespread blistering and can be fatal. Kartagener's syndrome is a triad of sinusitis, bronchitis and situs inversus caused by immotile cilia. **Down's syndrome** is a chromosomal disorder in which there are three copies of chromosome 21 in the karyotype. The wide range of symptoms of multiple sclerosis are caused by autoimmune destruction of myelin sheaths within the central nervous system. **Tay–Sachs disease** is a lysosomal storage disorder and is caused by a mutation in the gene coding for the alpha-chain of the enzyme hexosaminidase A.

Answers

57. 1 – F, 2 – B, 3 – H, 4 – C, 5 – E
58. T T T F T
59. T F F T T
60. 1 – F, 2 – I, 3 – E, 4 – H, 5 – C

SECTION 3

GENES TO PROTEINS

- NUCLEOTIDES AND NUCLEIC ACIDS (i) 72
- NUCLEOTIDES AND NUCLEIC ACIDS (ii) 74
- TRANSCRIPTION 76
- MATURATION OF EUKARYOTIC hnRNA INTO mRNA 78
- TRANSLATION (i) 80
- TRANSLATION (ii) 82
- THE HUMAN GENOME 84
- GENE EXPRESSION 86
- DNA PACKAGING AND REPLICATION 88
- RESTRICTION ENZYMES AND PLASMID VECTORS 90
- DNA ANALYSIS 92
- POLYMERASE CHAIN REACTION (PCR) 94
- DNA MUTATIONS (i) 96
- DNA MUTATIONS (ii) 98
- GENETIC TERMINOLOGY 100
- PEDIGREE CHART 102
- HEREDITARY DISORDERS (i) 102
- HEREDITARY DISORDERS (ii) 104
- GENETICS 104
- GENETIC COUNSELLING 106

SECTION 3 GENES TO PROTEINS

1. True or false? Concerning nucleotides

a. Nucleotides containing adenine, cytosine, guanine and thymine are found in ribonucleic acids
b. UTP, GTP and ATP act as stores of energy
c. They are glycosylated like proteins to modify function
d. They may be involved in cell signalling
e. Purine-containing nucleotides on each strand of double-stranded DNA base pair with each other

2. In double-stranded DNA

a. There are always equal amounts of guanine and cytosine nucleotides
b. Purine bases pair with pyrimidine bases
c. Uracil base pairs with adenine
d. Phosphodiester bonds link adjacent nucleotides
e. Hydrogen bonds are the major forces that maintain a double helix structure

3. Choose one term from options A–J which matches each description in the list 1–5

Options

A. Acid
B. Adenosine diphosphate
C. Adenosine monophosphate
D. Adenosine triphosphate (ATP)
E. Cyclic adenosine monophosphate
F. DeoxyATP
G. Deoxyribose
H. Phosphate
I. Purine
J. Ribose

1. The sugar component of a nucleotide found in ribonucleic acid
2. The group through which nucleotides are linked in nucleic acids
3. The nitrogen-containing part of nucleotides
4. An intracellular signalling molecule
5. A nucleotide molecule which is hydrolysed in cells to release energy

---

UTP, uridine triphosphate; GTP, guanosine triphosphate; ATP, adenosine triphosphate

EXPLANATION: NUCLEOTIDES AND NUCLEIC ACIDS (i)

The building blocks of nucleic acids are nucleotides (see diagram below). Nucleotides contain a pentose sugar linked to a nitrogen-containing base (via the 1′ carbon atom of the sugar) and to a phosphate group (via the 5′ carbon atom). Nucleosides contain just the pentose and the base. In nucleic acids, the building blocks are connected to each other via the phosphate groups; the phosphate group attached to the 5′ carbon atom of the pentose sugar of a nucleotide is linked to the hydroxyl group of the 3′ carbon atom of the pentose of another nucleotide.

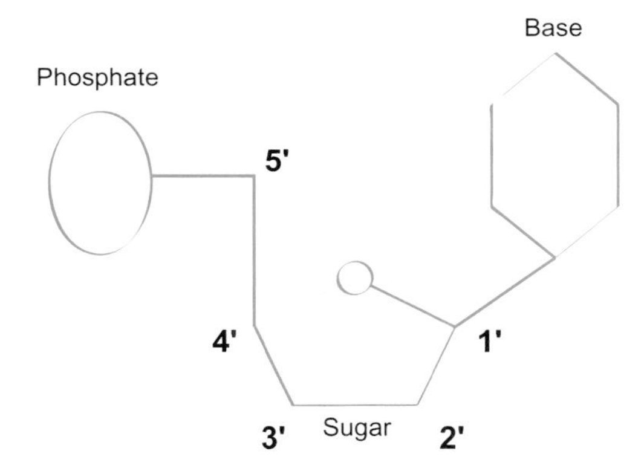

Though DNA and RNA are similar in structure there are subtle differences which greatly influence their function:

1. RNA is composed of nucleotides containing the pentose sugar **ribose**, whilst DNA is composed of nucleotides containing the pentose sugar **deoxyribose** (there is no hydroxyl group attached to the 2′ carbon atom).
2. The **purine** bases **adenine** and **guanine**, and the **pyrimidine** base **cytosine** are found in both DNA and RNA. The pyrimidine base **thymine** only occurs in DNA, whilst the pyrimidine base uracil only occurs in RNA.

The two strands of a **DNA double helix** interact with each other through **hydrogen bonds** between the bases. Purine bases bond with pyrimidine bases; the base adenine is able to 'base pair' with thymine and the base cytosine is able to bond with guanine.

Answers

1. F T F T F
2. T T F T T
3. 1 – J, 2 – H, 3 – I, 4 – E, 5 – D

4. Look at the molecule in the diagram below

a. Name the molecule
b. Name the component A in the diagram
c. Name the component B in the diagram
c. Name the component C in the diagram
e. What is this molecule used for?

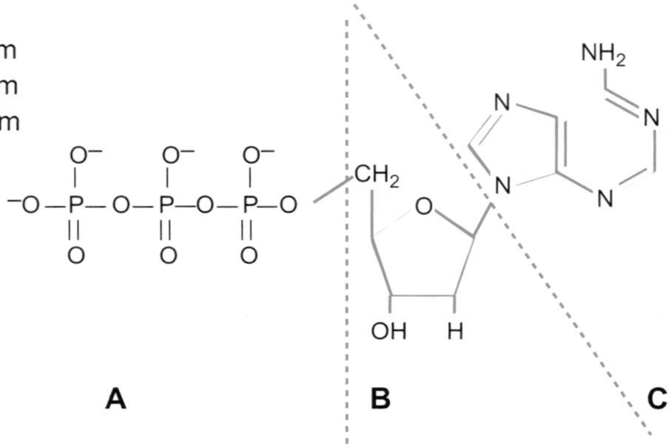

5. Theme – nucleic acids. From the options A–K, select the term that best matches the descriptions in questions 1–5

Options

A. Adenine
B. Covalent linkage
C. Cytosine
D. Guanine
E. Guanidine
F. Hydrogen bonds
G. Ionic interaction
H. Nucleosides
I. Nucleotides
J. Thymine
K. Uracil

1. The building blocks of nucleic acids
2. The force that stabilizes the DNA double helix
3. The base that pairs with cytosine in the DNA double helix
4. The base not found in DNA
5. The base that pairs with uracil in a DNA–RNA hybrid molecule

ATP, adenosine triphosphate; dATP, deoxyadenosine triphosphate; UTP, uridine triphosphate; GTP, guanosine triphosphate; cAMP, cyclic adenosine monophosphate; cGMP, cyclic guanosine monophosphate

EXPLANATION: NUCLEOTIDES AND NUCLEIC ACIDS (ii)

In double-stranded RNA molecules or in DNA–RNA hybrid molecules, uracil base pairs with adenine.

Nucleotides are biologically ubiquitous substances and have numerous other functions including acting as **stores of energies** (e.g. ATP, UTP, GTP). ATP (a ribonucleotide) is generated by the **glycolysis** pathway and by **oxidative phosphorylation**, and is a source of energy for many of the cell's functions. The terminal phosphate bond is highly labile and is hydrolysed easily to release usable energy. Note that dATP (which contains deoxyribose) is not used in the cell as a store of energy but is used as a substrate by DNA polymerase for DNA synthesis (4e). Nucleotides, such as cAMP and cGMP, are also used in intracellular signalling pathways. These intracellular signalling molecules are produced by the action of adenylyl cyclase and guanylyl cyclase on ATP and GTP respectively. In these cyclic nucleotides, the phosphate group of the nucleotide links the 5′ and 3′ carbon atoms within the molecule.

Though nucleotides contain a pentose residue, they are not glycosylated like proteins.

Answers

4. a. dATP; b. triphosphate group; c. deoxyribose; d. adenine; e. See explanation
5. 1 – I, 2 – F, 3 – D, 4 – K, 5 – A

6. Regarding transcription in eukaryotes

- **a.** It is the process by which DNA is copied into RNA
- **b.** It only occurs in the nucleus
- **c.** It requires the 60S ribosomal subunit
- **d.** It is initiated at the start (ATG) codon
- **e.** RNA polymerase II transcribes genes destined for translation

7. Concerning transcription in eukaryotes

- **a.** Ribosomal RNA is produced in the nucleoplasm by the action of RNA polymerase I
- **b.** Transfer RNA is produced in the nucleoplasm by the action of RNA polymerase III
- **c.** The sigma subunit of the RNA polymerase complex is required for the initiation of the RNA synthesis
- **d.** The Pribnow box is a region upstream of genes which is involved in the initiation of transcription
- **e.** Transcription is usually polycistronic

8. During transcription of messenger RNA in prokaryotes

- **a.** The DNA template is read from a 3′ to 5′ direction
- **b.** The RNA molecule is synthesized from a 3′ to 5′ direction
- **c.** The messenger RNA molecule is extensively modified during synthesis
- **d.** Palindromic sequences in the template can signal the termination of the process
- **e.** Translation can occur simultaneously

9. State three ways in which eukaryotic transcription differs from prokaryotic transcription

EXPLANATION: TRANSCRIPTION

Transcription, the process by which **DNA is copied into RNA** is mediated by **RNA polymerase** (composed of five subunits: two alpha, beta, beta prime and sigma). RNA polymerase assembles on **promoter regions** of genes upstream of the ATG translational start. This region contains specific sequences (e.g. the 'Hogness box' and the 'CAAT box' in eukaryotes and the 'Pribnow box' in prokaryotes), which allow the assembly of the polymerase. The sigma-subunit is required for recognition of the DNA and disassociates prior to elongation of the RNA molecule. The DNA template to be copied is **read from a 3′ to 5′ direction** and the RNA molecule is synthesized from a **5′ to 3′ direction**. Termination of transcription often occurs at palindromic sequences which generate a stem–loop secondary structure. Transcription in prokaryotes differs from that in eukaryotes as follows:

1. **Eukaryotic** transcription occurs in the **nucleus** (and **mitochondria**) whereas **prokaryotic** transcription occurs in the **cytoplasm**.
2. **Prokaryotes** have **one RNA polymerase** whilst **eukaryotes have three**:
   (a) RNA pol I transcribes ribosomal RNA (occurs in the nucleolus)
   (b) RNA pol II transcribes coding sequences (in the nucleoplasm)
   (c) RNA pol II transcribes transfer RNA (in the nucleoplasm).
3. **Prokaryotic** transcription is predominantly **polycistronic** (i.e. several genes are transcribed as one RNA molecule), whereas **eukaryotic** transcription is usually **monocistronic** (each gene is transcribed as a unique RNA molecule).
4. Prokaryotic transcription is coupled with translation. In eukaryotes, transcription and translation are distinct events (9).

Additionally, **eukaryotic** but not prokaryotic transcripts are **extensively modified** following transcription (see next section).

Answers

6. T F F F T
7. F T T F F
8. T F F T T
9. See explanation

10. Maturation of eukaryotic heteronuclear RNA can involve

- **a.** The addition of stretches of polyadenylic residues to the 3′ end of transcripts
- **b.** Editing of the RNA sequence
- **c.** The removal of exons
- **d.** The addition of a 5′–5′ guanosine cap
- **e.** The glycosylation of the ribose unit

11. Are the following correct associations between post-transcriptional modifications of heteronuclear RNA and the function of these modifications in eukaryotes?

- **a.** Capping of the 5′ end/promotes binding of transcript to ribosome
- **b.** Capping of the 5′ end/export of transcript to cytoplasm via the nuclear pore
- **c.** 3′ polyadenylation/promotes binding of transcript to ribosome
- **d.** 3′ polyadenylation/export of transcript to cytoplasm via the nuclear pore
- **e.** 3′ polyadenylation and 5′ capping/increases stability of transcript

12. The diagram below is a schematic representation of eukaryotic heteronuclear RNA. Choose one description from the list below which matches each of the numbered arrows (1–5) in the diagram

- **A.** Exon
- **B.** Intron
- **C.** Promoter
- **D.** Restriction enzyme site
- **E.** Site of addition of 7-methylguanosine residue
- **F.** Site of polyadenylation
- **G.** Splice acceptor site
- **H.** Splice donor site
- **I.** Start codon
- **J.** Stop codon

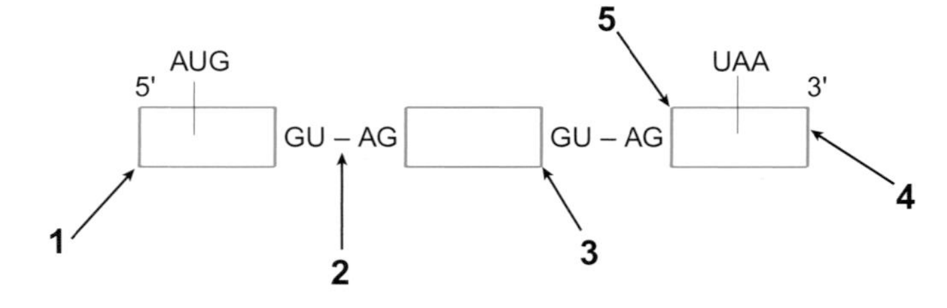

hnRNA, heteronuclear RNA; mRNA, messenger RNA; snRNA, small nuclear RNA

EXPLANATION: MATURATION OF EUKARYOTIC hnRNA INTO mRNA

Eukaryotic hnRNA is modified to generate mature **mRNA**. These modifications include:

1. The addition of a **7-methylguanosine cap** to the 5′ end of the transcript. This cap is added via a unique 5′–5′ linkage. Addition of the cap stabilizes the transcript, i.e. prevents rapid mRNA degradation, thus allowing translation to proceed. Furthermore the cap also promotes the binding of the transcript to the ribosome.
2. The 3′ end of the molecule is the site of **addition** of a long stretch of **adenine** containing nucleotides (polyadenylation). Polyadenylation stabilizes the transcript and allows the export of the mRNA from the nucleus into the cytoplasm via the nuclear pore.
3. Eukaryotic genes classically have **introns** introduced into the coding sequences (known as exons). These introns are copied during transcription and are **removed** from the RNA molecule subsequently by **RNA splicing**. The intron/exon boundary typically has a specific dinucleotide sequence. The 5′ end of the intron has the dinucleotide sequence GU and the 3′ end of the intron has the sequence AG. The 3′ end of the exon also has specific sequences (though less well conserved) and is known as the splice donor site. The 5′ end of the exon is known as the splice acceptor site. This splicing occurs on a 'spliceosome', a complex of several proteins and RNA molecules known as the small nuclear RNAs (snRNAs).
4. The **mRNA** molecule can also be '**edited**'. Editing is the addition or removal of specific bases, which allows the correct open reading frame to be generated. Significant editing occurs in the mitochondrial transcripts of human protozoan parasites such as *Leishmania* and *Trypanosoma*.

Answers

10. T T F T F
11. T F F T T
12. 1 – E, 2 – B, 3 – H, 4 – F, 5 – G

13. Translation in eukaryotes

- **a.** Occurs on the smooth endoplasmic recticulum
- **b.** Requires ribosomes
- **c.** Requires transfer RNAs
- **d.** Results in a polypeptide with methionine as the first amino acid
- **e.** Terminates at the AUG codon

14. Transfer RNAs

- **a.** Are exclusively found in eukaryotes
- **b.** Are longer than 1000 bases in length
- **c.** Usually adopt a 'cloverleaf' structure
- **d.** Have an invariable sequence (CCA) at their 3′ ends to which amino acids are attached
- **e.** Have an anticodon triplet which base pairs with messenger RNA

15. Theme – eukaryotic translation. Match one term from the list below for each statement 1–5

Options

- **A.** 30S subunit
- **B.** 50S subunit
- **C.** 60S subunit
- **D.** A site
- **E.** ATP
- **F.** Elongation factor Tu
- **G.** E site
- **H.** Guanosine triphosphate
- **I.** P site
- **J.** Small subunit
- **K.** Transfer RNA
- **L.** Uridine triphosphate
- **M.** Water

1. The portion of the ribosome which matches the correct transfer RNA to messenger RNA codons
2. The portion of the ribosome with peptidyltransferase activity that catalyses the formation of the peptide bond
3. The growing polypeptide chain remains attached to this molecule
4. The place on the ribosome that is able to accommodate an incoming aminoacyl transfer RNA
5. The hydrolysis of this molecule accelerates polypeptide synthesis

tRNA, transfer RNA; mRNA, messenger RNA; GTP, guanosine triphosphate; EFTu, elongation factor Tu

EXPLANATION: TRANSLATION (i)

Translation is mediated by **ribosomes** that are located in the **cytosol** and additionally in eukaryotes on **rough endoplasmic reticulum** (the 'roughness' conferred by ribosomes), and cannot proceed without tRNAs.

tRNAs, found in all living organisms, are **small RNA molecules** (<100 nucleotides in length) that adopt a cloverleaf tertiary structure. They have a conserved CCA sequence at their 3′ ends. Specific aminoacyl tRNA synthetases add amino acids to the 3′ hydroxyl group of this sequence. tRNAs also have an **anticodon** (three bases) that is able to pair with the appropriate **codon** of mRNA.

The eukaryotic and prokaryotic ribosomes are quite similar in design and function though they are different in mass; the prokaryotic ribosome is made of a **small 30S** and **large 50S** subunit whilst the eukaryotic ribosome is made of a **small 40S** and **large 60S** subunit. Translation is initiated when the small subunit of the ribosome binds to mRNA. This process is complex, requires additional proteins called initiation factors, and the expenditure of energy in the form of GTP hydrolysis. Before the mRNA is bound, a special initiator **tRNA**, charged with **methionine** and which recognizes the **AUG start codon**, is loaded onto the small ribosomal subunit. In prokaryotes this initiator tRNA is loaded with ***N*-formylmethionine**. The *N*-formylmethionine tRNA only associates with the small subunit of ribosomes at the initiation of transcription and is unable to enter the A site during chain elongation. Therefore, it cannot bind to AUG codons within the mRNA. The large subunit then binds to form an active ribosome.

The elongation of the polypeptide chain on ribosomes can be divided into several steps and different parts of the ribosomes perform distinct functions. The polypeptide chain remains attached to tRNA at all times. The ribosome contains three sites that are able to accommodate tRNAs: P, A and E sites. The P site contains the tRNA, to which is attached the growing polypeptide chain. The adjacent A site is able to accommodate incoming tRNAs charged with amino acids (aminoacyl tRNAs). Aminoacyl tRNAs enter the ribosomes in association with the elongation factor EFTu, to which is attached a molecule of GTP. Hydrolysis of the GTP occurs when the anticodon of the tRNA correctly pairs with the codon on mRNA. Hydrolysis of GTP results in a conformational change in EFTu which causes the release of the factor from the tRNA and the correct positioning of tRNA in the A site.

Answers

13. F T T T F
14. F F T T T
15. 1 – J, 2 – C, 3 – K, 4 – D, 5 – H

16. In prokaryotic translation, the formation of the initiation complex

a. Begins with the attachment of the 50S subunit to messenger RNA
b. Cannot occur without *N*-formylmethionine transfer RNA
c. Requires energy derived from the hydrolysis of ATP
d. Requires protein factors
e. Occurs on the AUG codon at the 5′ end of the messenger RNA

17. During the elongation stage of prokaryotic translation

a. Aminoacyl transfer RNAs base pair with the messenger RNA codon in the 'A site' of the ribosome
b. Binding of aminoacyl transfer RNA to ribosome is an energy-independent process
c. *N*-Formylmethionine transfer RNAs can bind to AUG codons within messenger RNA
d. The generation of peptide bonds is catalysed by ribosomal RNA
e. The growing polypeptide chain is attached to the 30S subunit

18. The following associations are correct regarding the drug and their effects on protein translation

a. Streptomycin/inhibits the initiation stage of translation in prokaryotic
b. Rifampicin/inhibits eukaryotic chain elongation
c. Cyclohexamide/inhibits eukaryotic peptidyltransferase activity
d. Puromycin/causes premature termination of the polypeptide chain
e. Tetracycline/inhibits binding of aminoacyl transfer RNA in prokaryotes

mRNA, messenger RNA; tRNA, transfer RNA

EXPLANATION: TRANSLATION (ii)

A **peptide bond** is then formed between the **growing polypeptide chain** and the **amino acid** in the A site. This bond formation is catalysed by a peptidyltransferase activity (intrinsic to the large subunit), and involves the transfer of the growing polypeptide chain to the N-terminus of the amino acid attached to the tRNA in the A site. This causes a series of conformational changes in the ribosomal structure that essentially moves the new tRNA (that is now attached to the polypeptide), into the P site. The old tRNA is shifted initially into the E site and is then finally ejected from the ribosome.

During these conformational changes, the **ribosome** also moves exactly **three bases** (one codon) along the mRNA molecule. This series of events is then repeated several times to complete the synthesis of the protein. The growing peptide chain always remains attached to tRNAs. It is never found to be attached to the ribosome subunits. These events occur quickly *in vivo*, with two amino acids being added to the polypeptide chain per second (eukaryotes). Prokaryote protein synthesis may be 10 times faster.

Translation **terminates** when **stop codons** are reached. The peptide is released and the ribosome disassociates to release the small and large subunits. Many antibiotics that are widely used in clinical practice inhibit translation (see question 18). Rifampicin inhibits the initiation of transcription in prokaryotes.

Answers

16. T T F T T
17. T T F T T
18. T F T T T

19. Regarding the human genome

a. It is organized into 46 chromosomes
b. It is the largest of any species thus far analysed
c. It is mostly composed of non-coding sequences
d. Seventy per cent of the genome is composed of repetitive sequences
e. It contains about 80 000 genes

20. The human genome (true or false?)

a. Is approximately the same size as that of *Clostridium* species
b. Is organized into 23 pairs of chromosomes
c. Contains 44 autosomes
d. Always contains an X chromosome
e. Always contains two copies of all genes.

21. Theme – the eukaryote genome. The following definitions refer to terms used in describing the genetic make-up of eukaryotic cells. Choose one term from the list below for each definition

Options

A. Diploid
B. Haploid
C. Heterozygote
D. Homozygote
E. Hybrid
F. Karyotype
G. Locus
H. Polyploidy
I. Triploid
J. Trisomy

1. A term describing an individual with two identical alleles for a gene on homologous chromosomes
2. A term describing an individual with two different alleles for a gene on homologous chromosomes
3. The number of chromosomes seen in normal gametes
4. A multiple of the haploid chromosome number (n), other than $2n$
5. The state where there are three copies of one chromosome in the genome

EXPLANATION: THE HUMAN GENOME

The human genome is about **3000 megabases** in size and is organized into **46 chromosomes** (22 pairs of autosomes and a pair of sex chromosomes; X and Y). The human genome always contains an **X chromosome** regardless of sex (i.e. XX or XY). Whilst in females there are always two copies of all genes, this is not true in males. In males, the **XY** configuration means that not all genes on these chromosomes are present in pairs. Coding sequences contribute approximately 3 per cent of the human genome and there are an estimated **80 000 genes**. Non-repetitive, single copy, non-genic sequences comprise about 50 per cent of the genome. About 15 per cent of the genome is composed of repetitive sequences which include mini- and micro-satellite DNA and transposons. The function of these non-coding DNA sequences remains unknown. Whilst many plants (e.g. pea, maize and wheat) as well as some animals (e.g. mouse and fruitfly) have much larger genomes than humans, prokaryotic genomes (such as *Clostridium*) are much smaller. Prokaryotic genomes are generally organized into a single circular chromosome that is present in the cytoplasm. Viral genomes are smaller than prokaryotic and eukaryotic genomes and can be composed of either RNA or DNA.

Gametes (sperm and egg) have a **haploid** number of chromosomes, i.e. they have one copy of each chromosome. Somatic cells (cells other than gametes in an individual) have two copies of each chromosome (**diploid**). The term polypoid is used to refer to the presence of more than twice the haploid chromosome number, e.g. $3n$, $4n$, $6n$, etc. **Trisomy** is a chromosomal abnormality in which there are **three copies** of a particular chromosome in a genome instead of the normal two. This seen in **Down's syndrome**, where there are three copies of chromosome 21 (**trisomy 21**). In diploid cells, the same genes on homologous chromosomes may be identical (homozygous) or different (heterozygous). The individuals are then referred to as homozygotes or heterozygotes.

Answers

19. T F T F T
20. F T T T F
21. 1 – D, 2 – C, 3 – B, 4 – H, 5 – J

22. True or false? Concerning gene expression

- **a.** Euchromatin represents transcriptionally active DNA
- **b.** Methylation of cytosine residues within 'CpG' islands of DNA prevents transcription
- **c.** Once a sequence is transcribed it will inevitably be translated to give rise to a functional protein
- **d.** RNA polymerase II transcribes genes encoding ribosomal RNA
- **e.** RNA polymerase reads the template from a 5′ to 3′ direction

23. Gene expression (i.e. the generation of a functional protein from a gene) can be controlled at the level of

- **a.** DNA structure
- **b.** Transcription
- **c.** Messenger RNA modification
- **d.** Translation
- **e.** Protein modification

24. Regarding gene activation by a steroid hormone in eukaryotes

- **a.** The steroid hormone diffuses across the plasma membrane
- **b.** The hormone receptor is located in the cytoplasm
- **c.** The hormone–receptor complex disassociates in the nucleus to release free hormone
- **d.** The free hormone binds to upstream regulatory elements of genes
- **e.** A single steroid hormone can cause the activation of over 50 genes

mRNA, messenger RNA; bp, base pair

EXPLANATION: GENE EXPRESSION

Heterochromatin is **tightly packaged DNA** that is not being actively transcribed. DNA that is being transcribed is unwound and disassociated from **histone** proteins and is known as **euchromatin**. mRNA is produced by DNA-dependent RNA polymerases which read the template from a 3′ to 5′ direction. mRNA is synthesized from a 5′ to 3′ direction.

Gene expression is controlled at several different levels including:

- At the level of **transcription**: some genes are not transcribed at all whilst others are transcribed at different levels. The DNA sequences around the promoter and upstream and downstream of the gene all have an effect on the rate of transcription. Different genes are transcribed by different RNA polymerases, e.g. RNA polymerase I transcribes ribosomal RNA genes (see earlier). Methylation at the C residues of 'CpG' islands also prevents transcription.
- At the level of **mRNA stability**: stability of mRNA is determined by post-transcriptional modification of the mRNA. Some mRNA molecules may be destined for destruction before translation.
- At the level of **translation** and **post-translationally** (by post-translational modification of the proteins).

Steroid hormones exert their effect by affecting gene expression. They are hydrophobic and hence readily **cross the plasma membrane** and penetrate the cell where they bind to receptors located within the **cytoplasm**. Once bound to these receptors, they enter the nucleus and are maintained as a complex. This complex then attaches to the 'hormone response elements' upstream of target genes. Typically the hormone response element is a 15-bp sequence specific to a particular steroid hormone–receptor complex. Each hormone–receptor activates some 50–100 genes and hence induces a large-scale change in the biochemical properties of the cell.

Answers

22. T T F F F
23. T T T T T
24. T T F F T

25. Concerning DNA structure and packing

- **a.** There are 10 base pairs per complete helical turn
- **b.** DNA binding proteins bind to the minor groove of the double helix
- **c.** Telomers are repetitive regions at the end of chromosomes
- **d.** Nucleosomes contain histone proteins
- **e.** Histone 4 links adjacent nucleosomes

26. The initiation of eukaryotic genome replication

- **a.** Occurs at specific DNA sequences
- **b.** Occurs at a single place on each chromosome
- **c.** Results in the emergence of two replication forks that progress in opposite directions
- **d.** Occurs in the M phase of the cell cycle
- **e.** Is controlled by checkpoints in the cell cycle

27. Regarding eukaryotic DNA polymerases

- **a.** They synthesize DNA in a 5′ to 3′ direction
- **b.** Read the template from a 3′ to 5′ direction
- **c.** DNA polymerase alpha has an exonuclease activity
- **d.** DNA polymerase delta has an intrinsic activity that unwinds DNA
- **e.** DNA polymerase gamma replicates mitochondrial DNA

28. Eukaryotic Okazaki fragments

- **a.** Are shorter than those in prokaryotes
- **b.** Require DNA polymerase alpha for their synthesis
- **c.** Require RNA primers for their synthesis
- **d.** Are generated during leading strand synthesis
- **e.** Are ligated together during DNA replication

29. Are the following correct associations between the named enzyme/protein and their function in eukaryotic DNA replication?

- **a.** DNA polymerase delta/proof reading activity
- **b.** Replication factor A/binds and stabilizes single-stranded DNA
- **c.** DNA polymerase alpha/leading strand synthesis
- **d.** Topoisomerase/winding and unwinding DNA
- **e.** Proliferating cell nuclear antigen/initiates RNA primer synthesis

---

PCNA, proliferating cell nuclear antigen

EXPLANATION: DNA PACKAGING AND REPLICATION

DNA is tightly packed in cells. Initially, lengths of **146 base pairs** of DNA are wound around an octomer of **histones** (two molecules each of H2A, H2B, H3 and H4) to form a **'beads on a string'** structure. These structures are called **nucleosomes** and this organization provides the first level of chromatin structure. DNA binding proteins bind in the major groove of the double helix, where information on the sequence of the DNA is available without separating the two strands. Adjacent nucleosomes are linked by the histone 1 protein and this structure is further coiled and folded into complex higher order structures which finally result in chromosomes.

The **initiation** of **genome replication** in eukaryotes occurs in the **S phase** of the cell cycle. It occurs at specific sequences on the DNA often referred to as origins of replication. There are numerous such sites on each chromosome and replication is initiated simultaneously at several points.

There are several types of DNA polymerases in eukaryotes. They **synthesize DNA in a 5′ to 3′ direction** and require a single-stranded template (the genomic DNA double helix is separated and maintained in a single-stranded configuration by protein factors). Unwinding and subsequent winding of the DNA is mediated by topisomerases. **DNA polymerase delta** synthesizes the **leading** strand. PCNA is an accessory factor to DNA polymerase delta that allows the polymerase to remain bound to the template DNA. **DNA polymerase alpha** is involved in the synthesis of the **lagging** strand. **Okazaki fragments** are generated during lagging strand DNA synthesis. In eukaryotes these fragments are shorter (~200 bases) than in prokaryotes (>2000 bases). DNA polymerase alpha produces a short RNA primer at specific points on the lagging strand and this is then elongated by DNA polymerase delta. Adjacent fragments are linked together by a DNA ligase after removal of the RNA primer and filling in. Some DNA polymerases (delta, beta and epsilon but not alpha) have exonuclease activities that are used to correct mistakes in the newly synthesized DNA strands.

Answers

25. T F T T F
26. T F T F T
27. T T F F T
28. T T T F T
29. T T F T F

30. Restriction endonucleases

a. Are bacterial enzymes
b. Recognize palindromic sequences such as 5′GGATCC3′
c. Cleave double-stranded DNA
d. Cleave double-stranded RNA
e. Cleave eukaryotic but not prokaryotic nucleic acids

31. Plasmid vectors used in the cloning of DNA

a. Are circular double-stranded DNA molecules
b. Have a selectable marker such as a drug resistance gene
c. Have a multiple cloning site which has recognition sequences for many restriction endonucleases
d. Do not require an origin of replication
e. Are located in the cytoplasm of host cells

32. In recombinant DNA technology, plasmid vectors

a. Can be used to amplify DNA
b. Can be used to clone RNA
c. May be used for the expression of proteins
d. Are only useful in prokaryotic systems
e. Like viruses, have to integrate into the host genome for propagation

33. From the options listed A–J, select an enzyme which matches the description given in each of the questions 1–5

A. Beta-lactamase
B. DNA polymerase
C. DNA ligase
D. *Eco*RI
E. Isomerase
F. Phosphodiesterase
G. Phosphofructokinase
H. Reverse transcriptase
I. RNA polymerase
J. Topoisomerase

1. Generates complementary DNA from messenger RNA
2. Joins two fragments of DNA
3. Cuts DNA at specific palindromic sites
4. Is used to generate a double-stranded molecule from complementary DNA
5. Uses uridine triphosphate as a substrate

cDNA, complementary DNA

EXPLANATION: RESTRICTION ENZYMES AND PLASMID VECTORS

Restriction enzymes are **bacterial enzymes** that attach to and cleave **palindromic sequences** within DNA. They have no effect on RNA. Such enzymes are thought to protect bacteria from viral infection by destroying viral DNA. Though DNA from any source can be cleaved by these enzymes, certain modifications of DNA, such as methylation of bases, may prevent enzyme activity. The cut ends of the DNA may be blunt ended, have a 5′ overhang or a 3′ overhang:

BLUNT END

5′—GGGCCC—3′	*Sma*I	5′—GGG	3′	+	5′	CCC—3′
3′—CCCGGG—5′	→	3′—CCC	5′	+	3′	GGG—3′

5′ OVERHANG

5′— GAATTC —3′	*Eco*RI	5′—G	3′	+	5′	AATTC—3′
3′— CTTAAG —5′	→	3′—CTTAA	5′	+	3′	G—5′

3′ OVERHANG

5′— GGTACC—3′	*Kpn*I	5′—GGTAC	3′	+	5′	C—3′
3′— CCATGG—5′	→	3′—C	5′	+	3′	CATGG—5′

These enzymes are very useful in the **cloning of DNA fragments** into plasmid vectors. **Plasmid vectors** are circular pieces of double-stranded DNA that are found in the cytoplasm of cells and which replicate independently of the genome. The plasmid vectors have an origin of replication (to ensure that the plasmid is replicated within the cell) and a selectable marker (e.g. an **ampicillin resistance gene**), so that cells containing them can be isolated and propagated and a unique restriction enzyme site into which the DNA fragment of interest can be inserted and cloned. The vector and the DNA of interest are cleaved with the appropriate restriction enzyme. The DNA of interest is then fused with the plasmid vector using an enzyme called DNA ligase. These plasmid vectors are introduced into and propagated in cells. Plasmids can be introduced into both eukaryotes and prokaryotes. RNA cannot be cloned and amplified using these vector systems. However, RNA can be converted to **cDNA** using an enzyme called **reverse transcriptase**. This cDNA is single stranded and DNA polymerase can be used to generate the double-stranded form, which can then be cloned using plasmid vectors.

Answers

30. T T T F F
31. T T T F T
32. T F T F F
33. 1 – H, 2 – C, 3 – D, 4 – B, 5 – I

34. The following statements concern agarose gel electrophoresis of DNA at pH 8.0. Choose one term from the list of options (A–L) below for the gaps in each statement

Options

A. Anode
B. Cathode
C. Charge
D. Ethidium bromide
E. Ethyl alcohol
F. Highest molecular weight
G. Lowest molecular weight
H. Methylene blue
I. Northern blotting
J. Size
K. Southern blotting
L. Western blotting

1. In agarose gels, the DNA molecules are separated according to 1
2. DNA migrates towards the 2
3. DNA with the 3 migrate furthest
4. The addition of 4 to the buffer allows visualization of DNA under ultraviolet light
5. The transfer of DNA separated by agarose gel electrophoresis to a solid support (e.g. nitrocellulose) is called 5

35. The following questions relate to techniques used in the diagnosis of genetic lesions in humans. Pick one term/phrase from the list (A–J) below which most accurately matches the description given in each question

Options

A. Agarose gel electrophoresis
B. Amniocentesis
C. Chorionic villus sampling
D. *In situ* hybridization
E. Karyotyping
F. Polymerase chain reaction
G. Restriction digest
H. Restriction fragment length polymorphism
I. Restriction site sequence
J. Tissue biopsy

1. A procedure used in prenatal diagnosis, where a small volume of the fluid surrounding the fetus is withdrawn by introducing a needle through the abdominal wall under ultrasonographic guidance
2. A procedure used in prenatal diagnosis, where a small amount of tissue is removed from the placenta transcervically or transabdominally under ultrasonographic guidance
3. The analysis of the genetic material from a biopsy, to determine the number of chromosomes in cells
4. A difference in DNA sequence between individuals that is recognizable when the DNA is digested with specific endonucleases such as *Eco*RI or *Bam*HI and separated by electrophoresis
5. The rapid amplification of short DNA fragments using oligonucleotide primers and a thermostable DNA polymerase

UV, ultraviolet; CVS, chorionic villus sampling; PCR, polyermase chain reaction

EXPLANATION: DNA ANALYSIS

In agarose **gel electrophoresis**, DNA loaded onto the gel **migrates** in an electric field. At pH 8.0, DNA is **negatively charged** and will move towards the **anode**. The rate of migration of the DNA in the electric field is dependent largely on size, with the smallest molecules migrating the fastest. Ethidium bromide, an agent that intercalates with DNA, is often added to the buffer during electrophoresis. This fluoresces under UV light, thus enabling the location of DNA within the gel matrix. After electrophoresis, the DNA within the matrix can be transferred to a **membrane** (e.g. nitrocellulose or nylon). This is called Southern blotting, and facilitates the identification of specific DNA molecules by hybridization with labelled probes.

Material for genetic analysis, can be obtained from the fetus using several techniques. These include **amniocentesis**, where a small volume of the amniotic fluid is removed using a needle introduced under ultrasound guidance. Amniotic fluid contains fetal cells which can be harvested and analysed. This technique can be used from 14 weeks onwards, when the risk to the fetus is relatively low (0.5 per cent chance of miscarriage). Amniocentesis can also be used to measure biochemical markers of disease present in the fluid (e.g. measurement of **alpha fetoprotein** for detecting neural tube defects). Samples can also be directly biopsied from the fetal trophoblast tissue (**chorionic villus sampling**). This can be done earlier than amniocentesis (from about 9–10 weeks gestation), but it carries a slightly higher risk of damage to the fetus. Nevertheless, CVS has a major advantage over amniocentesis in that it allows the results of diagnostic procedures to be available earlier. The material obtained from such procedures can be used to determine the karyotype of the cell and hence allow detection of abnormal chromosome number. The DNA obtained may also be **amplified by PCR** to determine specific gene defects (e.g. in cystic fibrosis).

In some genetic lesions, changes to the sequence of restriction enzyme sites may occur. The sequence may be deleted or may have a base change which no longer allows the enzyme to cut. These changes are known as **restriction fragment length polymorphisms** and may be used as markers for genetic defects.

Answers

34. 1 – J, 2 – A, 3 – G, 4 – D, 5 – K
35. 1 – B, 2 – C, 3 – E, 4 – H, 5 – F

36. Polymerase chain reaction – Short answer question

a. Explain the technique of polymerase chain reaction
b. Give two clinical uses of polymerase chain reaction

37. Polymerase chain reaction

a. Can be used to derive double-stranded DNA from single-stranded DNA
b. Theoretically results in a linear amplification of DNA
c. Requires the presence of oligonucleotide primers and a thermostable DNA polymerase
d. Requires at least 1 μg of template
e. Can be used in the prenatal diagnosis of inherited conditions such as cystic fibrosis

PCR, polymerase chain reaction; cDNA, complementary DNA; dNTP, deoxynucleotide triphosphate; HIV, human immunodeficiency virus

EXPLANATION: POLYMERASE CHAIN REACTION (PCR)

PCR is a technique that is used to exponentially **amplify** minute amounts (less than 1 ng) of either single- or double-stranded DNA (known as the template). RNA can also be amplified by this technique after its conversion to single-stranded DNA (known as cDNA). PCR requires the presence of two short, single-stranded DNA fragments that are complementary to the template at either end (known as **oligonucleotide primers**), a **DNA polymerase** that is stable at high temperatures and dNTPs. The following diagram shows the sequence of events that is involved in PCR. Double (or single)-stranded DNA is heated to a high temperature to denature the DNA (abolish base pairing) (Step 1). Primers anneal to the template as the temperature is lowered (Step 2).

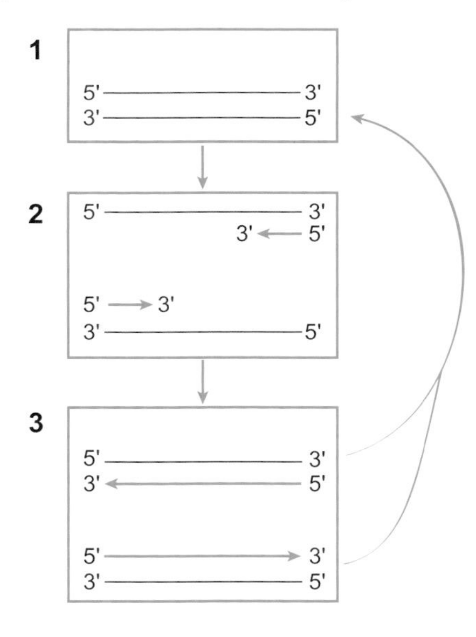

The primers are then **extended** by the thermostable DNA polymerase with the dNTPs as substrate. This then generates two copies of the original double-stranded template (Step 3). **The cycle is then repeated many times**, the net result of which is an exponential amplification of the template (36a).

Examples of the clinical application of PCR include (36b):

- The detection of genetic lesions, e.g. mutations in the cystic fibrosis genes or in alpha and beta globin genes in the thalassaemias.
- The detection of viral DNA/RNA, e.g. HIV. The strain type can also be determined.
- The detection of drug resistance of certain pathogens, e.g. in *Mycobacterium tuberculosis.*

M. tuberculosis grows very slowly *in vitro* and hence the conventional methods of detecting drug sensitivity may take months. PCR offers a rapid alternative to detect genes that will confer resistance to drugs.

Answers

36. See explanation
37. T F T F T

38. Regarding DNA mutations

- **a.** A point mutation that changes an adenine to a thymine residue is known as a transversion
- **b.** A point mutation that changes a cytosine to a guanine residue is known as a transition
- **c.** A deletion of 9 base pairs of DNA will result in a frameshift mutation
- **d.** Transversions can result in a frameshift mutation
- **e.** A nonsense mutation generates a premature stop codon

39. Insertional mutation of DNA

- **a.** Can involve 1 to several 1000s of bases
- **b.** Seldom occurs at repeated motifs
- **c.** Can be caused by transposons
- **d.** Can be caused by viruses
- **e.** Can be caused by *Staphylococcus aureus*

40. Theme – DNA mutations.

The diagram below shows some of the sequence of the open reading frame of a gene. The sequence begins at the start codon. The boxes (1–5) indicate the sites of possible mutations. Choose one term from the options list that is most descriptive of the mutations indicated in the following questions

Options

A. Deletion **B.** Duplication **C.** Frameshift **D.** Insertion **E.** Inversion
F. Missense **G.** Nonsense **H.** Substitution **I.** Transition **J.** Transversion

1. The replacement of A with G at point 1
2. The introduction of G at point 2
3. The replacement of C with A at point 3
4. The replacement of C with A at point 4
5. The introduction of the sequence AGT AGT AGT at point 5

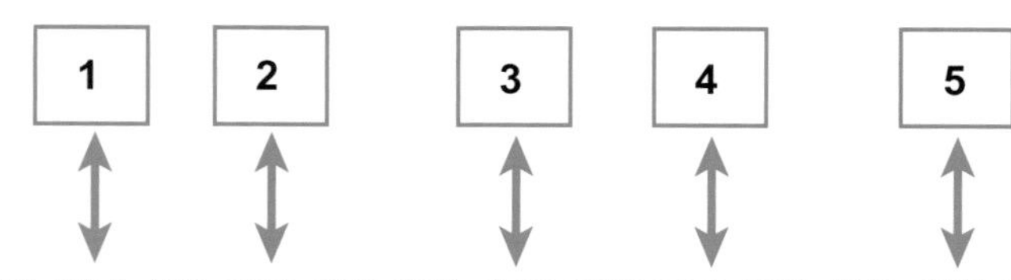

EXPLANATION: DNA MUTATIONS (i)

The **substitution** of a purine for a purine or a pyrimidine for a pyrimidine residue is called a **transition mutation** (A and G are purines and C and T are pyrimidines). When such transition mutations occur at the third base of a codon, there is almost never any effect on the protein sequence. The only exception is when the codon TGG (for tryptophan) is mutated to TGA (a stop codon). The generation of stop codons by mutation is called a nonsense mutation. Other stop codons are TAA and TAG.

Transversion mutations occur when a **purine** is substituted for a **pyrimidine** residue. Such mutations often change the amino acid specified by the codon.

Insertion or deletion of a length of DNA sequence where the length is not a multiple of three will cause a **frameshift mutation** (i.e. the open reading frame of the gene will be changed downstream of the mutation site). Frameshift mutations do not occur where there is insertion or deletion of entire codons (i.e. deletion/insertion of base pairs in multiples of three). Insertions and deletions into DNA that cause mutations often occur at repeated sequences, often by slippage of the DNA polymerase. Insertional mutations can also be caused by the integration of transposons and viral DNA into the host genome. Bacteria such as *Staphylococcus aureus* are extracellular and hence do not integrate their genomes into host sequences.

Answers

38. T F F F T
39. T F T T F
40. 1 – I, 2 – C, 3 – J, 4 – G, 5 – D

41. DNA mutations can be caused by

a. Replication slippage during DNA synthesis
b. Ultraviolet radiation
c. Ethidium bromide
d. Puromycin
e. Reactive oxygen species

42. The following associations are correct regarding DNA mutations

a. Ethidium bromide/insertion mutation
b. Ultraviolet radiation/covalent linking of adjacent thymine residues
c. Alkylating agent/point mutation
d. Deaminating agent/triplet repeat expansion
e. 5-Bromouracil/deletion mutation

43. Which of the following definitions of biological agents are correct?

a. Mutagen – causes mutations in the sequence of DNA
b. Clastogen – causes fragmentation of chromosomes
c. Mitogen – causes neoplastic transformation of eukaryotic cells
d. Oncogen – induces tumour formation
e. Teratogen – causes developmental abnormalities

UV, ultraviolet

EXPLANATION: DNA MUTATIONS (ii)

DNA can be **damaged** in a wide variety of ways, including as a result of the effects of **free radicals**, **chemicals** and **electromagnetic radiation**. Alkylating agents, deaminating agents and base analogues such as 5-bromouracil all cause **point mutations** of DNA. Intercalating agents such as ethidium bromide are flat molecules which slip between the bases in the double helix and are associated with **insertional mutations**. UV light causes covalent linking of adjacent bases, which are subsequently removed to give rise to **deletion mutations**. Puromycin is an antibiotic that inhibits translation and has no effect on DNA. Mutations can also be introduced during DNA replication, where the replication machinery may stutter (thereby adding extra bases: insertional mutation) or when the polymerase 'slips' and fails to copy a sequence of DNA (deletion mutations).

Biological agents which cause cell dysfunction can be classed according to their mechanism of action:

1. MUTAGEN – causes mutations in the sequence of DNA
2. CLASTOGEN – causes fragmentation of chromosomes
3. MITOGEN – causes resting cells to enter the cell cycle and hence cell division. The proliferating cells are normal in structure and function
4. ONCOGEN – induces tumour formation
5. TERATOGEN – causes developmental abnormalities

Answers

41. T T T F T
42. T T T F F
43. T T F T T

44. The following questions refer to some common definitions used in genetics. Choose one term from the options below that best fits each description in 1–10

Options

A. Autosomal recessive
B. Congenital
C. Linked
D. Monozygotic
E. Mosaicsm
F. Penetrance
G. Polygeneic
H. Wild type
I. X-linked recessive
J. Y-linked

1. A gene that may be carried by females, but that is expressed only in males carrying one copy of the gene
2. A term used to describe a gene that is present at any location other than the X and Y chromosomes and where males and females may be carriers
3. A term used to describe a genetic trait that only manifests in males and never in females
4. A term used to describe the pattern of inheritance that is determined by many genes located at different loci
5. A term used to describe a normal allele
6. A term used to describe a set of twins that are genetically identical
7. A term used to describe two genes that tend to be transmitted together
8. A term to describe the degree to which a dominant allele is expressed in a population of heterozygotes
9. A term to describe the presence in an individual of two different cell types derived from the same zygote but differing genetically
10. A term used to describe a trait that is present at birth, but which does not necessarily have a genetic basis

45. True or false? An autosomal dominant gene

a. Will exhibit a phenotype in heterozygotes
b. Will be transmitted to all progeny from a homozygote parent
c. Cannot be carried on the Y chromosomes
d. May be carried on the X chromosome
e. Cannot be transmitted from males to males

46. An X-linked recessive gene

a. Can be transmitted from females to females
b. Can be transmitted from males to males
c. Will always be transmitted from father to daughter
d. Will only exhibit its phenotype in males
e. Is always lethal in males

EXPLANATION: GENETIC TERMINOLOGY

In **X-linked conditions**, all **males** are affected as they only carry one X chromosome. X-linked dominant traits will be expressed in female heterozygotes.

Genetically identical (monozygotic) twins are derived from the same egg and sperm fusion product. Dizygotic twins will contain different genetic material as they are derived from the fertilization of different eggs at the same time.

Some genes do not always exert a phenotype in all individuals. **Penetrance** is expressed as a percentage. A penetrance of 70 per cent for a gene implies that the 70 per cent of a population that carry the gene will have the phenotype. Expressivity is the term used to describe the extent to which a genetic defect is expressed in an individual.

Autosomal dominant genes by definition are carried on autosomes. Sex chromosomes (X and Y) do not carry them. They can be transmitted from either parent to offspring regardless of sex. Because they are dominant, they will have a phenotypic effect in heterozygotes. Offspring of homozygotes for the gene will all have a copy of the gene.

X-linked traits cannot be transmitted from father to son; the father has to donate the Y chromosome. Conversely, a father carrying an X-linked trait will always pass on the gene to daughters; the father has to donate the X chromosome in this case. X-linked recessive traits can exert their phenotypes in both males and in females. It is more likely in males because they only carry one X chromosome. However a female can acquire two X-linked recessive genes and thus the phenotype will be manifest. X-linked recessive genes are not always lethal (e.g. haemophilia A and Duchenne's muscular dystrophy).

Answers

44. 1 – I, 2 – A, 3 – J, 4 – G, 5 – H, 6 – D, 7 – C, 8 – F, 9 – E, 10 – B
45. T T T F F
46. T F T F F

47. The symbols in the diagram below are used to describe individuals in a pedigree chart (family tree). Match one symbol with each of the following descriptions

1. A female affected by the genetic lesion
2. Carrier of an X-linked recessive trait
3. A deceased male
4. A male who is heterozygote for the trait
5. Monozygotic twins

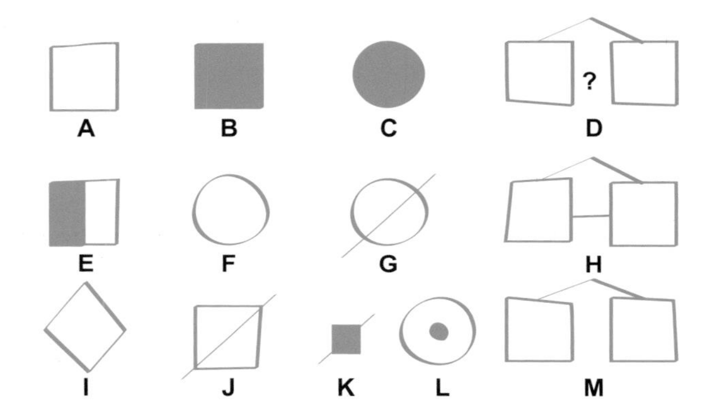

48. Regarding hereditary disorders. Select disorders from the options below that are caused by the genetic lesions listed as 1–5

Options

A. Achondroplasia
B. Angelman's syndrome
C. Cystic fibrosis
D. Down's syndrome
E. Haemophilia A
F. Huntington's chorea
G. Kleinfelter's syndrome
H. Patau's syndrome
I. Prader–Willi's syndrome
J. Sickle cell disease
K. Turner's syndrome
L. Type I diabetes mellitus

1. Point mutation in a gene
2. Trisomy
3. X-linked recessive gene
4. Triplet repeat expansion
5. Loss of a whole chromosome

EXPLANATION: PEDIGREE CHART

In pedigree charts males are represented by squares and females by circles. A diamond shape indicates that the sex is unknown. Completely shaded shapes indicate that the individual is affected by the genetic trait under consideration. Shading of half the shape indicates that the individual is a heterozygote for the trait. A line through the shape indicates that the individual is dead. Prenatal death is indicated by the symbol K in the diagram opposite. Females who are carriers of an X-linked trait are represented by the symbol L.

Twins are represented by the symbols D, H and M in the diagram opposite. D indicates that twins are of unknown zygosity. Monozygotic twins are indicated by H and dizygotic twins by the symbol in M.

EXPLANATION: HEREDITARY DISORDERS (i)

Genetic diseases can arise from a number of lesions including:

1. POINT MUTATIONS A **single base change** within a gene can have dramatic effects on the function of the gene. The gene may not be expressed at all or, if expressed, the resulting protein may be dysfunctional hence causing disease. The classic example of a disease resulting from such a mutation is **sickle cell** disease (see later).
2. ABNORMAL CHROMOSOME NUMBER Though the vast number of embryos with an abnormal chromosome number do not survive, there are some notable exceptions:

   - **Down's syndrome (trisomy 21)**. This condition is associated with a characteristic flat facial profile, abundant neck skin, cardiac malformations and mental retardation.
   - **Edward's syndrome (trisomy 18)**. This condition is also associated with a characteristic facial appearance, rocker bottom feet, flexion of the limbs and mental retardation. Mean survival is 10 months.
   - **Patau's syndrome (trisomy 13)**. There is characteristic facial appearance with a small head and eyes. This condition is associated with polycystic kidneys, cardiac malformations, cleft lip/palate and mental retardation.
   - **Turner's syndrome (45XO)** (a total of 45 chromosomes with a loss of an X chromosome). Girls with this condition are typically short, with a webbed neck and a broad chest. The ovaries are rudimentary and puberty may not occur. It is associated with coarctation of the arota and cardac malformations.
   - **Kleinfelter's syndrom**e (polysomic sex chromosomes, i.e. **XXY** or **XXXY**). This is the commonest cause of hypogonadism in males and is usually detected at puberty where it may present with learning difficulties, gynaecomastia, sparse facial hair or the failure of sexual maturity. This syndrome is associated with infertility.

3. TRIPLET REPEAT EXPANSION Expansions in non-coding repeated regions involved in gene regulation may give rise to disease states. For example, **Huntington's disease** is caused by the expansion of a CAG repeat in exon I. **Fragile X syndrome** is caused by expansion of CTG repeats in the 3′ untranslated region; **Friedreich's ataxia** by expansions of GAA repeats in intron 1.

Answers

47. 1 – C, 2 – L, 3 – J, 4 – E, 5 – H
48. 1 – J, 2 – D, 3 – E, 4 – F, 5 – K

49. The genetic lesion in sickle cell disease

- **a.** Is caused by a point mutation in the beta globin gene
- **b.** Can be caused by any one of a number of distinct mutations of globin genes
- **c.** Is due to a frameshift mutation
- **d.** Is X-linked
- **e.** Is maintained in the population because of advantage to heterozygotes

50. Regarding cystic fibrosis (true or false?)

- **a.** It is an autosomal recessive condition
- **b.** 1 in 25 in the population is a carrier
- **c.** Steatorrhoea is a feature
- **d.** The liver is unaffected
- **e.** Males are infertile

51. The cystic fibrosis gene

- **a.** Is found on chromosome 7
- **b.** Encodes a ligand-gated sodium channel
- **c.** ΔF508 mutation is the most common mutation seen in cystic fibrosis
- **d.** Mutations can be screened for using PCR antenatally
- **e.** Defect is transmitted to progeny exclusively by females

52. Case study

John is a one-year-old child. His mother, Angie, is aware that several members of her family have a blood disorder. Angie has two other children – one boy who is affected and one girl who is not. Angie also has two brothers, one of whom is affected and one sister who is not. Neither of her parents nor her husband are affected by the condition. She has no information on her grandparents. Angie's sister has four children, two girls and two boys. Both the boys, but neither of the girls, have manifested the symptoms of the condition.

- **a.** Draw a family tree to show the pattern of inheritance of the condition in this family
- **b.** What is the mode of inheritance of the disease?
- **c.** What is the probability that John has inherited the condition?
- **d.** Name two conditions that have this mode of inheritance

CFTR, cystic fibrosis transmembrane conductance regulator; PCR, polymerase chain reaction

EXPLANATION: HEREDITARY DISORDERS (ii)

The gene lesion in **sickle cell disease** is a single point mutation in the beta globin gene. This results in an amino acid substitution in the globin protein (glutamine to valine at position 6). The resulting mutant protein polymerizes at low oxygen tensions causing erythrocytes to adopt the classic sickle shape. It is an autosomal recessive trait and is maintained at high frequency in Africa because it provides some protection in heterozygotes against malaria.

Cystic fibrosis is an **autosomal recessive** condition. The gene (located on the long arm of chromosome 7) encodes the CFTR protein. This is an integral membrane protein that serves as a **Cl^- channel**. Some 70 per cent of mutations result in the deletion of a phenylalanine residue at position 508 (the ΔF508 mutation), and molecular biological techniques such as PCR and Southern blotting can be used to identify such mutations. The CFTR gene is expressed in all epithelial cells; mutations result in reduced secretion of Cl^- ions and hence reduced amount of water**.** The net result is the secretion of viscous mucus, which blocks airways in the lungs and ducts in the pancreas. This leads to airway disease and disease of the pancreas that eventually leads to malabsorption and steatorrhoea. The bile ducts in the liver are also often blocked, leading to liver disease. In males, the vas deferens is affected, leading to loss of fertility.

EXPLANATION: GENETICS

In the case study described, the mode of inheritance is **X-linked recessive**: only males are affected, suggesting a sex-linked pattern (see pedigree chart below). Not all males are affected, therefore the condition cannot be caused by a dominant gene on the Y chromosome. Furthermore, it cannot be caused by a dominant gene on the X chromosome, as females carrying this genes would manifest symptoms. The only explanation is that it is an X-linked recessive trait. Hence only males inheriting the faulty gene on the X chromosome will display the phenotype. Females who are **heterozygotes** for the gene are **carriers**. Note that females can also display the phenotype if they carry two X chromosomes, both of which contain defective genes (52b).

The chance that John has inherited the condition is 1:2. John's mother must be a heterozygote (i.e. has one X chromosome carrying the recessive gene and the other chromosome carrying the normal gene), as one of her sons is affected (52c).

X-linked recessive conditions include colour blindness, haemophilia, muscular dystrophy and X-linked deafness syndrome (52d).

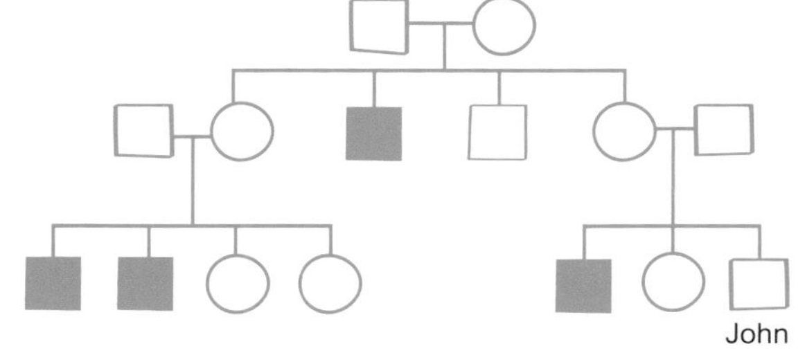

Answers

49. T F F F T
50. T T T F T
51. T F T T F
52. See explanation and diagram

53. Case study

You are consulted at the genetic counselling clinic by Paul and Nicola who tell you that they are planning to have a child. They are worried that their child might be affected by cystic fibrosis as Paul's brother Andrew died from the complications of cystic fibrosis at age 25. You know that cystic fibrosis is an autosomal recessive trait with an allele frequency of 1:43 in your local population.

- **a.** What is the chance that Paul is a carrier for the cystic fibrosis gene?
- **b.** What is the chance that Nicola is a carrier for the cystic fibrosis gene?
- **c.** What are the chances their planned child might be affected by the disease?

EXPLANATION: GENETIC COUNSELLING

In the case study opposite, both of Paul's parents must have been carriers of the cystic fibrosis gene. Therefore, the chance that Paul is a carrier can be worked out as follows. In this scenario, we will represent **the cystic fibrosis gene as 'c'** and the **normal gene as 'C'**. Both of Paul's parents have the genotype **Cc**. The possible genotypes of their children are as shown below:

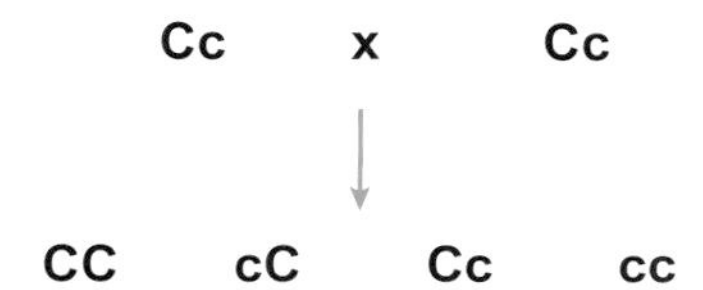

Clearly, Paul cannot have the cc genotype as he does not have cystic fibrosis. Therefore, the chance that he is a carrier is **2:3** (i.e. Paul has the genotype cC or Cc) (53a).

The cystic fibrosis gene occurs with a frequency of **1:25** in the UK. The cystic fibrosis gene occurs at a **frequency of 1:43** in the population referred to in this question. Therefore the normal gene has a frequency of 42/43. The chance that Nicola is a carrier can be worked out as follows. To be a carrier, Nicola must have the genotype cC or Cc. The chances of cC or Cc = $(1/43 \times 42/43) + (42/43 \times 1/43) = 2(1/43 \times 42/43) = 0.045$ or 1:22. Thus there is a 1:22 chance that Nicola is a carrier of the defective gene (53b).

The chances that a child will be affected by the disease can be calculated as follows. If both Nicola and Paul are carriers of the cystic fibrosis gene, then there is 1:4 chance that the child will be homozygous recessive, i.e. have a cc genotype (see diagram above). Therefore the chance of a child being affected is: $1/4 \times 1/22 \times 2/3 = 1/132$ (53c).

Genetic screening for cystic fibrosis is available. Both Paul and Nicola may be screened to determine if either are a carrier for the gene. If one of them is not a carrier, then any child that they might have will not be affected by the disease. If they are both carriers, then there is 1:4 risk that their child will be affected . **Prenatal genetic testing** may be used to determine the genotype of the fetus. This knowledge may be useful to the parents in making decisions regarding the continuation of the pregnancy.

Answers

53. See explanation

SECTION 4

PROTEIN STRUCTURE AND FUNCTION

- AMINO ACIDS (i) 110
- AMINO ACIDS (ii) 112
- AMINO ACID SIDE CHAINS 114
- PROTEIN STRUCTURE (i) 116
- PROTEIN STRUCTURE (ii) 118
- PROTEIN STRUCTURE (iii) 120
- POST-TRANSLATIONAL PROTEIN MODIFICATION 122
- PEPTIDES 124
- BLOOD PROTEINS AND AMINO ACIDS 126
- HAEMOGLOBIN (i) 128
- HAEMOGLOBIN (ii) 130
- EXTRACELLULAR MATRIX MOLECULES 132
- COLLAGENS 134
- CYTOSKELETAL PROTEINS 136
- ANTIBODIES 136
- INSULIN SECRETION 138
- DISEASES ARISING FROM DEFECTS IN PROTEINS 140
- PROTEIN ANALYSIS 142
- OEDEMA AND ASCITES 144

SECTION 4 PROTEIN STRUCTURE AND FUNCTION

1. Regarding the structure of amino acids

a. All amino acids have an amino group
b. Some amino acids do not have a carboxyl group
c. Most amino acids exist in solution as zwitterions at pH 7
d. The side groups are attached to the alpha carbon atom
e. Only the L-enantiomers of amino acids are found in proteins

2. Amino acids

a. Are the building blocks of proteins
b. Can act as neurotransmitters
c. Can be precursors of neurotransmitters
d. Can be used for the synthesis of membrane phospholipids
e. Can be used in gluconeogenesis

3. For each definition (1–5), choose an amino acid from the list A–L below

Options

A. Alanine
B. Asparagine
C. Aspartate
D. Glutamine
E. Cysteine
F. Glycine
G. Ornithine
H. Phenylalanine
I. Proline
J. Histidine
K. Serine
L. Tyrosine

1. An essential amino acid in humans
2. An amino acid not usually found in proteins
3. An amino acid with an acidic side chain
4. An amino acid that is the precursor of adrenaline
5. An amino acid that prevents alpha helix formation

4. The following amino acids have positively charged side chains at neutral pH

a. Lysine
b. Glycine
c. Arginine
d. Leucine
e. Histidine

EXPLANATION: AMINO ACIDS (i)

All **amino acids** have an **amino** and **carboxyl group** attached to the alpha carbon atom. The side groups (R in the diagram below) are also attached to the alpha carbon. The **alpha carbon** of all amino acids, with the exception of glycine, is therefore **chiral** and only the L-enantiomers are used to build proteins *in vivo*. In solution, the carboxyl group loses a proton and becomes negatively charged, whilst the amino group gains a proton and becomes positively charged. Such molecules that contain both a positive and negative charge are called **zwitterions**.

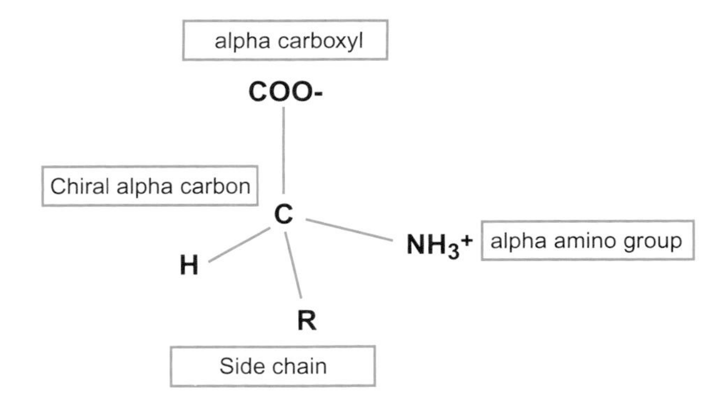

The properties of the side groups of amino acids are important, as they impact crucially on the structure and function of protein.

- Amino acids with basic side chains: lysine, arginine, histidine
- Amino acids with acidic side chains: aspartic acid, glutamic acid
- Amino acids with uncharged polar side chains: asparagine, glutamine, serine, threonine, tyrosine
- Amino acids with non-polar side chains: alanine, valine, leucine, isoleucine, proline, phenylalanine, methionine, tryptophan, glycine, cysteine

The side chains of **arginine** and **lysine** ionize and have a **positive** charge at neutral pH. Histidine has a pK of 6.5 and will exist as a mixture of positively and uncharged molecules at pH 7. The **acidic** side groups of **glutamate** and **aspartate** ionize to give rise to a COO^- group at neutral pH. Many side chains (those of serine, threonine, asparagine and glutamine) do not ionize. However, they are still polar. The side chain group of tyrosine may ionize at high pH values, or in special microenvironments within a folded protein.

Answers

1. T F T T T
2. T T T T T
3. 1 – H, 2 – G, 3 – C, 4 – L, 5 – I
4. T F T F T

5. The side chains of the following amino acids do not ionize in aqueous solution

a. Aromatic hydroxyl group of tyrosine
b. The hydroxyl group of serine
c. The thioalcohol group of cysteine
d. The amide group of glutamine
e. The carboxyl group of glutamate

6. The following amino acids have aromatic side chains

a. Phenylalanine **b.** Tyrosine **c.** Cysteine **d.** Histidine **e.** Tryptophan

7. Select one amino acid from the options A–J, which best matches the statements in each question, 1–5

Options

A. Alanine **B.** Arginine **C.** Glutamate **D.** Glycine **E.** Isoleucine
F. Leucine **G.** Methionine **H.** Phenylalanine **I.** Threonine **J.** Valine

1. An amino acid with a side chain that has a negative charge at pH 7
2. An amino acid with a side chain with a positive charge at pH 7
3. An amino acid with a non-ionizing polar side chain
4. An amino acid with a side chain with the smallest possible size
5. An amino acid with a side chain that contains sulphur

8. From the amino acid listed as options A–J, select the amino amino acid which contains the side group indicated in each question 1–5

Options

A. Arginine **B.** Cysteine **C.** Glutamate
D. Glutamine **E.** Glycine **F.** Histidine
G. Methionine **H.** Tryptophan **I.** Tyrosine
J. None of the above

1. A thiol group
2. A thioether group
3. A carboxylic acid group
4. An amide group
5. A phenol group

EXPLANATION: AMINO ACIDS (ii)

Phenylalanine, **tyrosine** and **tryptophan** are **aromatic amino acids** and contain a **benzene ring**.

Two amino acids commonly found in proteins contain **sulphur** in their side chains. These are **cysteine** which contains a thiol group ($-CH_2-SH$) and methionine which contains a thioether group (CH_3-S-CH_2-). No amino acid contains a sulphate group.

The amino acids which contain amides in their side chains are glutamine and asparagine ($-CO-NH_2$).

Glycine has the smallest side group of all. It contains a single hydrogen as the side group.

The ring structure of the proline side chain prevents the formation of alpha helices and introduces kinks into the peptide chain.

Amino acids are very versatile molecules and are used in a variety of ways in biological systems. Apart from being the **building blocks** of proteins, they are used in a number of ways including the synthesis of membrane phospholipids (e.g. phosphatidylserine), as **neurotransmitters** (e.g. glutamate) or as the **precursors** of neuro-transmitters (e.g. tyrosine is the precursor of both dopamine and noradrenaline). Amino acids can also be deaminated and then converted into glucose within cells.

Although there are many different possible amino acids, **only 20 occur commonly** in proteins. Other amino acids, like ornithine, are intermediates in metabolic pathways.

There are **nine essential amino acids** that cannot be synthesized by the human body. These amino acids must be absorbed from the diet and include: threonine, lysine, methionine, valine, phenylalanine, leucine, tryptophan and isoleucine. Histidine is probably not essential in adults, but may be essential for growing children.

Answers

5. F T F T F
6. T T F F T
7. 1 – C, 2 – B, 3 – I, 4 – D, 5 – G
8. 1 – B, 2 – G, 3 – C, 4 – D, 5 – I

9. Theme – titration of amino acids. Look at the pH titration curve shown below for an amino acid. Pick the most appropriate answer from the list (A–O) below for each question

Options

- **A.** Glycine
- **B.** Histidine
- **C.** Glutamate
- **D.** Asparagine
- **E.** Lysine
- **F.** Aspartate
- **G.** 4
- **H.** 5
- **I.** 6.5
- **J.** 8
- **K.** 10
- **L.** 12
- **M.** pH 3.5–4.8
- **N.** pH 5.5–7.5
- **O.** pH 8.5–11.5

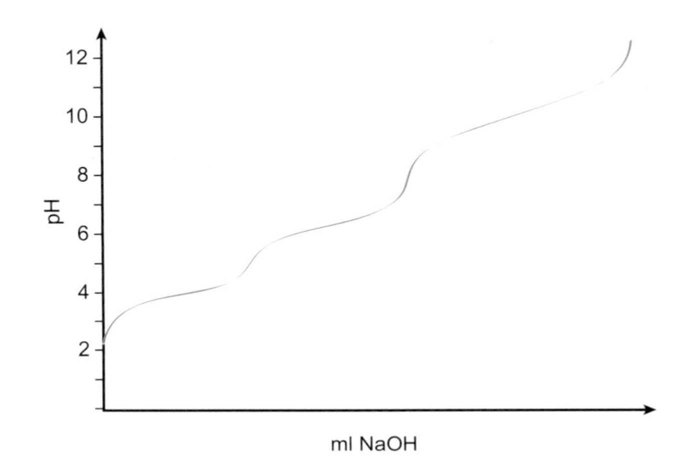

1. What is the pK value of the alpha carboxyl group of this amino acid?
2. What is the pK value of the alpha amino group of this amino acid?
3. What is the pK value of the side chain group of this amino acid?
4. What is this amino acid?
5. What is the buffering range of this amino acid side chain?

10. The pK of the side chains of the following amino acids is greater than that of histidine

a. Arginine **b.** Tyrosine **c.** Glutamate **d.** Lysine **e.** Aspartate

11. The side chains of the following amino acids within a protein may be involved in the indicated function

- **a.** Histidine: buffering
- **b.** Serine: hydrophobic interaction
- **c.** Glutamate: ionic bond formation
- **d.** Cysteine: disulphide bridge formation
- **e.** Valine: hydrophobic interactions

EXPLANATION: AMINO ACID SIDE CHAINS

The **pH at which exactly half of the molecules with a charged group are in the ionized form is called the pK of that group**. The pK of the alpha carboxyl group of most amino acids is in the range of 4, whilst the pK of the alpha amine group is in the region of 10. The pK of the side chain group of different amino acids is variable. The side chains of **basic** residues (arginine, lysine) and tyrosine have **high pK values**, whilst the **acidic** side chains of glutamate and aspartate have **lower ones**. The pK_a of the histidine side chain is 6.5 and hence it is able to perform as a good buffer under physiological conditions where the pH is maintained between 7.35 and 7.45.

The **side chains** of amino acids give proteins their **specific properties**. Hydrophobic interactions are mediated by the hydrophobic side chains of amino acids like valine, isoleucine, phenylalanine and leucine (tyrosine may have a charged side chain). Ionic bonds can form between charged side chains like glutamate or aspartate and lysine or arginine. Two cysteine residues can become oxidized to form a disulphide bridge (this is an example of a covalent bond).

Answers

9. 1 – G, 2 – K, 3 – I, 4 – B, 5 – N
10. T T F T F
11. T F T T T

12. During the formation of a peptide bond between isoleucine and valine, where valine is the C-terminal residue

- **a.** The bond forms between the carboxyl group of isoleucine and the amino group of valine
- **b.** The bond forms between the side chains of the amino acids
- **c.** Bond formation results in the generation of a molecule of water
- **d.** Isoleucine will retain its carboxyl group
- **e.** Valine will lose its side chain group

13. During protein folding

- **a.** The folding conformation is largely determined by the primary structure of the protein
- **b.** Hydrophobic side chains are usually arranged on the outside of the protein molecule
- **c.** Hydrogen bonds play an important role in maintaining structure
- **d.** Peptide bonds interact with one another to form hydrogen bonds
- **e.** Disulphide bonds are crucial for maintaining structure in the cytosol

14. The following statements concern the forces involved in maintaining the structure of proteins. Choose one term from the list (A–I) below that fits each description

Options

A. 180 degrees	**B.** 45 degrees	**C.** 60 degrees
D. 90 degrees	**E.** Covalent bond	**F.** Hydrogen bond
G. Hydrophobic interaction	**H.** Ionic bond	**I.** Van der Waals attraction

1. A bond that depends purely on electrostatic interaction between two oppositely charged atoms
2. A bond where an electropositive atom is shared between two electronegative atoms
3. The weakest form of bond
4. Hydrogen bonds are strongest when all the atoms involved are arranged at this angle relative to each other
5. A bond that is not broken by the process of heat denaturation

15. Protein structure is maintained by

- **a.** Hydrogen bonds
- **b.** Electrostatic interactions
- **c.** Van der Waals forces
- **d.** Covalent bonds between glutamate side chain groups
- **e.** Hydrophobic interactions

EXPLANATION: PROTEIN STRUCTURE (i)

Peptide bonds form between the **alpha carboxyl** groups and the **alpha amino groups** of amino acids. Peptide bond formation does not involve the side groups of amino acids. This reaction is a condensation reaction as it generates a molecule of water. After bond formation, one of the amino acids will have a free alpha amino group and this end of the peptide is called the N-terminus. Similarly, at the other end of the peptide, another amino acid will have a free alpha carboxyl group and this end is known as the C-terminus. The amino acid sequence of a protein (the primary structure) in general determines how a protein is going to fold. Hydrophilic residues tend to attract water and are usually arranged near the surface of the folded structure of globular proteins. In contrast, hydrophobic residues interact with each other and are found buried within the protein. Protein structure is maintained by many types of bonds:

1. HYDROGEN BONDS The peptide bonds are polar and interact with each other and with other polar side groups to form hydrogen bonds. Hydrogen bonds are special types of bonds where the hydrogen atom, which is electropositive, is shared between two electronegative atoms; one of the electronegative atoms is the atom to which the hydrogen is covalently linked. This type of bond is highly directional and is strongest when all three atoms are in a straight line, i.e. at an angle of 180 degrees relative to each other.
2. ELECTROSTATIC INTERACTIONS (ionic bonds) Form between positive and negative side group chains.
3. VAN DER WAALS FORCES These are weak and occur when atoms are close together. The sheer number of van der Waals interactions present in any protein make it a very important player in maintaining protein structure.
4. HYDROPHOBIC INTERACTIONS These occur within the protein core of globular proteins and are a driving force for protein folding.
5. COVALENT BONDS Disulphide bridges can form between the side chains of cysteine. They are important in maintaining the structure of secreted or surface proteins. In the cytosol however, disulphide bonds rarely form. The cytosol contains high concentrations of –SH–reducing substances that break disulphide bonds.

The relative strengths of these bonds are:

Covalent bond > ionic bond > hydrogen bond > van der Waals interaction

Answers

12. T F T F F
13. T F T T F
14. 1 – H, 2 – F, 3 – I, 4 – A, 5 – E
15. T T T F T

16. From the list of options (A–M) choose the answer which best fits the statements in each question 1–5

Options

- **A.** Active site
- **B.** Catalytic group
- **C.** Dimeric structure
- **D.** Hydrophobic
- **E.** Hydrophilic
- **F.** Primary structure
- **G.** Prosthetic group
- **H.** Quaternary structure
- **I.** Secondary structure
- **J.** Side chain group
- **K.** Substrate group
- **L.** Super secondary structure
- **M.** Trimeric structure

1. A term which describes the level of organization in a protein composed of more than one polypeptide chain
2. A term that describes the linear sequence of amino acids in a protein
3. A term that describes the haem part of myoglobin
4. A term that describes the structural arrangement of alpha helices and beta sheets into regular motifs within proteins
5. A term which describes the region of the papain enzyme that binds to and cleaves the substrate

17. True or false? Alpha helices in proteins

- **a.** Are left-hand helices
- **b.** Require proline for formation
- **c.** Are stabilized by hydrogen bonds
- **d.** contain 3.6 amino acid residues per complete turn of helix
- **e.** Are generally absent in globular proteins

18. Regarding beta sheets in proteins

- **a.** Formation requires an extended polypeptide chain
- **b.** The antiparallel beta sheet is more stable and rigid than the parallel beta sheet
- **c.** Hydrogen bonds between the side chains of amino acids hold the sheet together
- **d.** Covalent bonds are required additionally to hydrogen bonds to maintain structure
- **e.** Beta sheets are found in the core of many globular proteins

EXPLANATION: PROTEIN STRUCTURE (ii)

The **amino acid sequence** of a protein is known as the **primary structure**. The configuration adopted by a local region within a protein is called secondary structure and includes the alpha helices and beta sheets.

The **alpha helix** is a **secondary protein structure** present in many fibrous (e.g. keratin) and globular proteins (e.g. globins). The helix is **right-handed** and when groups of alpha helices come together in fibrous proteins they form a left-handed superhelical structure. The formation of the helix is spontaneous and is stabilized by **hydrogen bonds** between the carbonyl and the amide nitrogen of peptide bonds spaced four residues apart. There are 3.6 residues per complete turn of the helix. Amino acids such as glutamate, isoleucine and methionine are favoured in the formation of this secondary structure, and their side chains are arranged on the outside of the helix.

In contrast, amino acids such as glycine and proline disrupt the formation of a helical structure. The coiled-coil structure is a highly stable, stiff structure that is formed by the winding together of two alpha helices. The alpha helices are usually identical, run in the same direction and contain repeating non-polar hydrophobic amino acid residues. Often they are large, and serve as the building blocks of larger fibrous structures, e.g. the thick filament of muscle cells. The helices interact with each other via these hydrophobic residues. Short coiled-coil structures can serve as motifs that allow two proteins to dimerize (e.g. in several gene regulatory proteins).

Beta sheets form when extended polypeptide chains fold back and forth along each other. The folding may result in a **parallel** beta sheet (where the direction of the adjacent polypeptides is the same; direction is determined by N- and C-terminals of each polypeptide), or an **antiparallel** beta sheet (where adjacent polypeptides run in opposite directions). The structure is maintained by hydrogen bonds between adjacent peptide bonds. The side chains do not contribute to bond formation. The hydrogen bonds in the antiparallel sheet are arranged in parallel to each other and give rise to a more stable and rigid structure. Covalent bonds are not involved in maintaining structure.

The **tertiary structure** describes how the protein is finally folded into a conformation (e.g. **globular** or **fibrous**). The association of several folded polypeptide chains by non-covalent interaction into a larger structure is the quaternary structure (e.g. dimeric, tetrameric, etc.). The overall folding of the protein allows the formation of important functional domains (e.g. the catalytic or active sites), which are able to bind to substrate and perform a catalytic reaction. The side chain groups that perform the catalytic function in these sites are called **catalytic groups**.

Answers

16. 1 – H, 2 – F, 3 – G, 4 – L, 5 – A
17. F F T T F
18. T T F F T

19. Super secondary structure

a. Describe what is meant by super secondary structure. Give two examples of super secondary structure and an example of the proteins in which they occur
b. What is the functional significance of super secondary structures?

20. The coiled-coil structure

a. Is relatively unstable
b. Is generated from two interacting alpha helices and a beta sheet
c. Has two alpha helices that are usually identical in primary structure
d. Has two alpha helices with repeating hydrophilic residues that interact to generate the structure
e. Can act as a motif which allows protein dimerization

21. The following are correct associations

a. Alpha helix/secondary protein structure
b. Beta barrel/super secondary protein structure
c. Beta sheet/tertiary protein structure
d. Beta-alpha-beta fold/tertiary protein structure
e. Dimeric/quaternary protein structure

22. Theme – protein structure. Choose one protein from the list (A–J) below that has the structural properties indicated in each question

Options

A. Actin filament
B. Alpha tubulin
C. Collagen
D. Elastin
E. Haemoglobin
F. Insulin
G. Keratin
H. Oxytocin
I. Pyruvate kinase domain 1
J. Vasopressin

1. A globular protein with significant amounts of alpha helix
2. A fibrous protein with significant amounts of alpha helix
3. A protein with a triple helical structure
4. A protein with a significant number of hydrophobic residues organized into random coils
5. A protein with a beta barrel super secondary structure

EXPLANATION: PROTEIN STRUCTURE (iii)

Analysis of the tertiary structure of proteins has shown that secondary structures are often arranged into patterns. Certain patterns of organization of the secondary structures have been recognized and are known as super secondary structures. Such super secondary structures may be found in many, often unrelated, proteins. Examples of super secondary structures include the following.

1. BETA-ALPHA-BETA FOLD The beta sheets are arranged within the protein in a classical twisted structure and are connected by alpha helices. The twisted beta sheets are arranged within the centre of the protein with the alpha helices arranged on the outside. Examples of proteins with this type of structure include flavodoxin, lactate dehydrogenase domain 1 and phosphoglycerate kinase domain 2.
2. ALPHA HELIX BUNDLE The alpha helices, joined by short segments of polypeptide, fold back on themselves to create a bundle which may sometimes be globular. Such structures are often found in the interior of a protein. Examples of proteins with alpha helix bundles include myoglobin, haemoglobin, lysosome and cytochrome C.
3. BETA BARREL The beta sheets are arranged in a barrel shape with each sheet interconnected by an alpha helix. Again the alpha helices are arranged on the outside of the structure. Examples of proteins with beta barrels include triose phosphate isomerase and pyruvate kinase domain 1.

Super secondary structure motifs may be important in stabilizing the overall conformation of a protein (i.e. may have a structural role) or may form distinct functional domains within proteins. For example, the helix-turn-helix motif is a super secondary structure that allows several transcription factors to bind to DNA. The immunoglobulin domain structure, which is found in many different proteins, allowing the grouping of these proteins into a superfamily, is a beta barrel super secondary structure **(19)**.

Answers

19. See explanation
20. F F T F T
21. T T F F T
22. 1 – E, 2 – G, 3 – C, 4 – D, 5 – I

23. Give four examples of post-translational modifications of proteins. Indicate where they are likely to occur. Give a function for each type of modification

24. The structure, activity and function of proteins can be modified by

- **a.** Glycosylation
- **b.** Phosphorylation
- **c.** Membrane potential
- **d.** pH
- **e.** Temperature

25. Glycosylation of proteins

- **a.** Takes place in the endoplasmic reticulum
- **b.** Takes place in the Golgi apparatus
- **c.** Can occur at glutamate residues
- **d.** Can occur at asparagine residues
- **e.** Can occur at serine residues

26. *N*-linked glycosylation of proteins

- **a.** Requires the translocation of the peptide into lysosomes
- **b.** Involves the transfer of oligosaccharides from dolichol pyrophosphate
- **c.** Targets the protein for transfer to the nucleus
- **d.** Only occurs at asparagine residues in the sequence asparagine–leucine–serine
- **e.** Targets the protein for export to the cell surface

27. Choose the amino acid from the list (A–J) that would most likely be the target for the post-translational modifications indicated in each question 1–5

Options

- **A.** Alanine
- **B.** Asparagine
- **C.** Cysteine
- **D.** Glutamic acid
- **E.** Isoleucine
- **F.** Leucine
- **G.** Methionine
- **H.** Proline
- **I.** Tryptophan
- **J.** Tyrosine

1. *N*-glycosylation
2. Vitamin K-dependent carboxylation
3. Disulphide bond formation
4. Phosphorylation by protein kinases
5. Vitamin C-dependent hydroxylation

EXPLANATION: POST-TRANSLATIONAL PROTEIN MODIFICATION

Protein structure, function and activity are all modified constantly in a cell by many mechanisms. Most enzymes have a narrow pH and temperature range in which they are active. Furthermore, changes in the membrane potential of a cell can also change the structure of the protein which allows it to perform a certain function (e.g. activation of a voltage-gated Na^+ channel by membrane depolarization). The most common protein modifications are as follows (23):

1. REMOVAL OF SIGNAL SEQUENCE A signal sequence is required for the targeting of proteins to certain organelles. For example, a hydrophobic sequence at the N-terminus allows the protein to enter the endoplasmic reticulum and a sequence of positively charged residues is required for import into the nucleus. These signals are often removed in the organelles by specific peptidases and this allows maturation of the protein.
2. DISULPHIDE BRIDGE FORMATION The thiol groups of cysteine residues can become oxidized to form covalent disulphide bonds. Disulphide bridges are formed in the endoplasmic recticulum and may be either intramolecular or intermolecular, serving to hold different polypeptide chains together.
3. GLYCOSYLATION Sugar motifs can be added to specific residue side chains. The asparagines in the sequence Asn–X–Ser/Thr (where X = any amino acid except glycine or proline) can become glycosylated. This is known as *N*-glycosylation and occurs in the endoplasmic reticulum. In the lumen of the endoplasmic reticulum a branched oligosaccharide is assembled on dolichol pyrophosphate, and is transferred to specific asparagine residues of target proteins by the enzyme oligosaccharide protein transferase. The hydroxyl groups of serine and threonine residues may also become modified by the addition of carbohydrate moieties. This type of modification is known as *O*-glycosylation and occurs in the **Golgi complex**. These types of modifications are important structurally and also allow the protein to be targeted to specific locations (e.g. to the cell membrane or to the extracellular matrix). In the Golgi, phosphorylation of the mannose residues of glycoproteins prevents export from the cell and instead allows the proteins to be shunted to lysosomes, where they are either destroyed or function as lysosomal proteins.
4. MODIFICATION OF AMINO ACIDS Amino acid side chains can also become modified by a variety of substances apart from glycosylation. The side chains of specific serine or threonine or tyrosine residues may become **phosphorylated** by specific kinases in the cytoplasm. The phosphorylation of these residues may activate/deactivate the function of these proteins. Lysine and proline residues may also become **hydroxylated**, as in collagen by a vitamin C-dependent reaction. Here these residues are involved in creating bonds which keep the overall structure of the protein. Yet other amino acid side chains may become carboxylated by a vitamin K-dependent (e.g. the carboxyl side chain of glutamate in clotting factors) or lipidated (to anchor proteins to membrane surfaces).
5. PROTEOLYTIC CLEAVAGE Proteins can be cleaved to generate active molecules (e.g. proinsulin to insulin).

Answers

23. See explanation
24. T T T T T
25. T F F T T
26. F T F T T
27. 1 – B, 2 – D, 3 – C, 4 – J, 5 – H

28. Regarding peptides. Look at the dipeptide below: From the list of options (A–K) select the appropriate answer for each question

Options

- **A.** +1
- **B.** +2
- **C.** 0
- **D.** −1
- **E.** −2
- **F.** Arginine
- **G.** Aspartate
- **H.** Glutamate
- **I.** Glutamine
- **J.** Glycine
- **K.** Lysine

1. Name the amino acid at the N-terminus
2. Name the amino acid at the C-terminus
3. What is the charge of the dipeptide at pH 2?
4. What is the charge of the dipeptide at pH 7?
5. What is the charge of the dipeptide at pH 11?

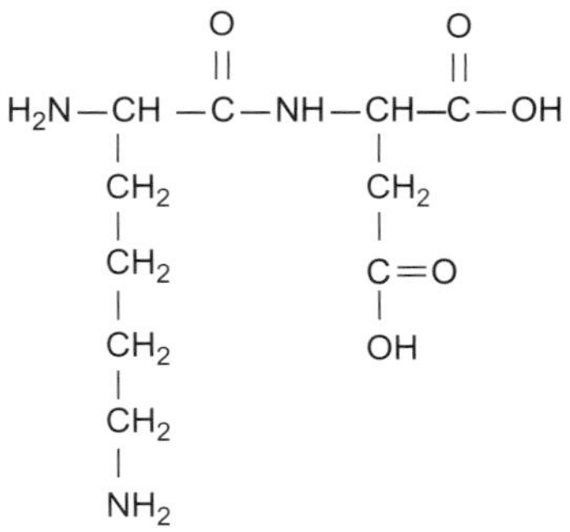

```
              O             O
              ||            ||
H2N—CH  —C—NH—CH—C—OH
     |              |
    CH2            CH2
     |              |
    CH2            C=O
     |              |
    CH2            OH
     |
    CH2
     |
    NH2
```

29. Peptides

- **a.** Draw the structure of the following hexapeptide: (N-terminus) aspartate–cysteine–tyrosine–proline–lysine–cysteine (C-terminus)
- **b.** Name two common post-translational modifications that may happen to this peptide. Indicate which residues will be involved in the modification
- **c.** What would the net charge of this peptide be at physiological pH?

EXPLANATION: PEPTIDES

In question 29, the N-terminal amino acid is lysine and the C-terminal residue is aspartate. At **pH 2**, all carboxyl groups will be in the **COOH form**. The amino groups will become **protonated**, i.e. become NH_3^+. Therefore the total charge would be +2. At **pH 7** the carboxy groups will be deprotonated (**COO^- form**) and the amino groups will remain **protonated**. Therefore the net charge is 0. At **pH 11** all the above functional groups will lose **protons**. Therefore the net charge will be −2 (because of the two resulting COO^- groups).

The hexapeptide aspartate–cysteine–tyrosine–proline–lysine–cysteine has the following structure (29a):

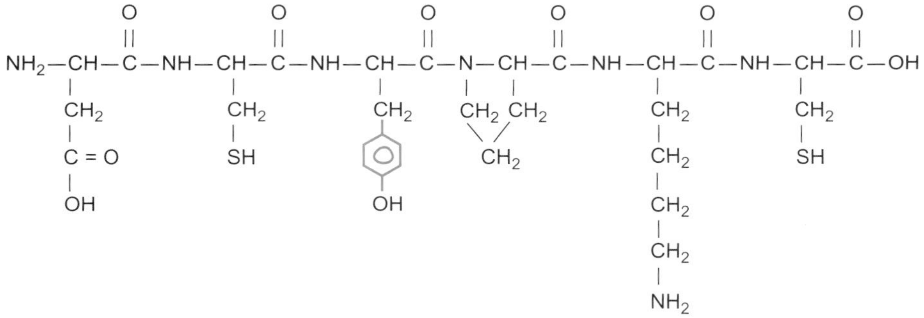

```
           O           O           O        O          O          O
           ||          ||          ||       ||         ||         ||
NH2—CH—C—NH—CH—C—NH—CH—C—N—CH—C—NH—CH—C—NH—CH—C—OH
    |           |           |    |  |          |          |
    CH2         CH2         CH2  CH2 CH2       CH2        CH2
    |           |           |     \ /          |          |
    C = O       SH       (phenyl) CH2          CH2        SH
    |                       |                  |
    OH                      OH                 CH2
                                               |
                                               CH2
                                               |
                                               NH2
```

The following modifications are possible:

- **Disulphide bridges** can form between the two cysteine residues. These can be either intra- or interpeptide disulphide bridges
- The **tyrosine residue** may become **phosphorylated** by a kinase enzyme using ATP
- The **proline** or **lysine** residues may become **hydroxylated** as in collagen (29b)

At physiological pH, the structure of the molecule would be:

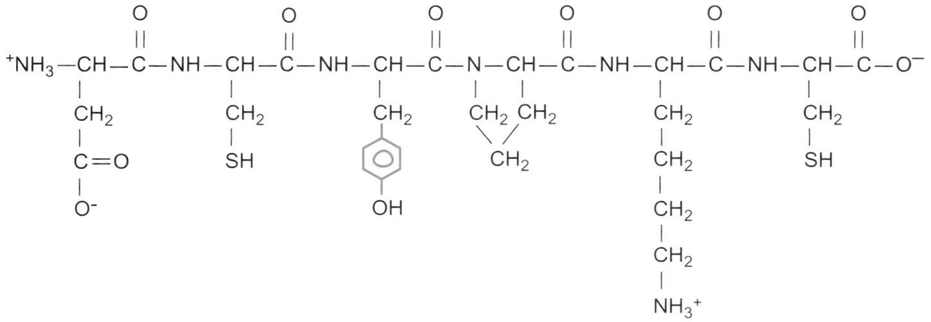

```
            O           O           O        O          O          O
            ||          ||          ||       ||         ||         ||
+NH3—CH—C—NH—CH—C—NH—CH—C—N—CH—C—NH—CH—C—NH—CH—C—O−
     |           |           |    |  |          |          |
     CH2         CH2         CH2  CH2 CH2       CH2        CH2
     |           |           |     \ /          |          |
     C=O         SH       (phenyl) CH2          CH2        SH
     |                       |                  |
     O-                      OH                 CH2
                                                |
                                                CH2
                                                |
                                                NH3+
```

Therefore the net charge would be 0 (4c).

Answers

28. 1 – K, 2 – G, 3 – B, 4 – C, 5 – E

29. See explanation

30. The following proteins are involved in transport of substances in the blood

- **a.** Transferrin
- **b.** Haemoglobin
- **c.** Albumin
- **d.** Immunoglobulin
- **e.** Interferon gamma

31. True or false? Human albumin

- **a.** Is a monomeric globular protein
- **b.** Is synthesized by the kidneys
- **c.** Is less abundant than immunoglobulins in plasma
- **d.** Can be used by cells as a supply of amino acids
- **e.** Deficiency can result in oedema

32. Choose one protein found in blood from the list below which is involved in the process indicated in each question 1–5

Options

- **A.** Albumin
- **B.** Alkaline phosphatase
- **C.** Ceruloplasmin
- **D.** Ferritin
- **E.** Fibrinogen
- **F.** Gamma globulins
- **G.** Interferon gamma
- **H.** Lipoprotein lipase
- **I.** Microglobulins
- **J.** Transferrin

1. Maintaining colloid osmotic pressure of plasma
2. Coagulation of blood
3. Transport of copper
4. Transport of iron in plasma
5. Humoral response to a viral or bacterial infection

33. Regarding the plasma amino acid concentration

- **a.** The normal plasma amino acid concentration is 35–65 mg/dL
- **b.** Plasma amino acid concentration rises sharply post-prandially
- **c.** The plasma amino acid concentration is dependent solely on absorption from the gastrointestinal tract
- **d.** Adrenocortical glucocorticoid hormone increases plasma amino acid concentration
- **e.** Growth hormone lowers plasma amino acid concentrations

EXPLANATION: BLOOD PROTEINS AND AMINO ACIDS

Plasma contains a variety of proteins, each one of which has a specific function. The major protein component of plasma is **albumin**. Albumin (molecular weight 65 kDa) is a monomeric globular protein synthesized by the liver. Some 10–15 g is synthesized per day and synthesis can be upregulated or downregulated as required. Albumin performs a variety of functions:

- **Transport of substances** Many substances are hydrophobic and are not soluble in plasma (e.g. **cortisol**, **thyroid hormone**, **bilirubin** and **fatty acids**). These substances can be transported bound to proteins such as albumin. Drugs such as **warfarin** and **aspirin** as well as some ions (e.g. Ca^{2+} and zinc) are also transported bound to albumin.
- **Maintaining the osmotic pressure of plasma**. Albumin contributes about 80 per cent of the total effective plasma osmotic pressure. In this way filtered fluid is allowed to be reabsorbed back into the plasma from the interstitial space. Albumin deficiency results in tissue oedema.
- **Nutrition**: albumin can be degraded by endothelial and other cells as a **source** of **amino acids**.

Other major protein components of plasma include **fibrinogen** (which is capable of polymerizing into long fibrin threads during blood coagulation) and **immunoglobulins** (which are involved in antibody responses to a immunological challenge). The blood also contains a number of specialized proteins that are involved in transport of various substances, e.g. transferrin (iron) and ceruloplasmin (copper).

In vivo, the plasma amino acid concentration is maintained at a constant level and is independent of absorption from the gastrointestinal tract. Post-prandial plasma concentrations of amino acids do not rise sharply for two reasons. First, protein digestion and absorption typically occur over a few hours and only small quantities of amino acids enter the plasma per unit time. Secondly, after entering the blood, excess amino acids are rapidly absorbed by cells throughout the body (especially of the liver and kidney) and are stored as proteins. The **concentration** of **plasma amino acids** can be altered by several hormones, including **insulin** and **growth hormone** (which encourage uptake into cells) and by **glucocorticoid hormone** (increases plasma concentrations).

Answers

30. T T T F F
31. T F F T T
32. 1 – A, 2 – E, 3 – C, 4 – J, 5 – F
33. T F F T T

34. The following subunits of haemoglobin are produced at the indicated developmental stage in humans

- **a.** Alpha-globin: adult
- **b.** Alpha-globin: fetal
- **c.** Beta-globin: adult
- **d.** Gamma-globin: adult
- **e.** Delta-globin: fetal

35. Haemoglobin A

- **a.** Is a multimeric protein
- **b.** Contains two alpha- and two beta-subunits
- **c.** Contains one haem group around which the subunits are arranged
- **d.** Can bind to two molecules of oxygen
- **e.** Is the major type of haemoglobin found in adults

36. Haemoglobin A

- **a.** Shows allosteric binding properties towards O_2
- **b.** Binds tightly to carbon monoxide
- **c.** Has a higher affinity for carbon monoxide than for oxygen
- **d.** Has a lower affinity for oxygen than fetal haemoglobin
- **e.** Has a higher affinity for oxygen than myoglobin

Loan Receipt
Liverpool John Moores University
Library Services

Borrower Name: Davies,Lois
Borrower ID: *********1113**

Cell and molecular biology /
31111011453410
Due Date: 05/05/2016 23:59

Total Items: 1
25/04/2016 19:32

Please keep your receipt in case of dispute.

EXPLANATION: HAEMOGLOBIN (i)

The systemic requirements of oxygen far exceed the amount that can dissolve in plasma and consequently an effective delivery system is required. Three per cent of oxygen in arterial blood is dissolved in the plasma and the remaining 97 per cent is found in association with haemoglobin. Different types of haemoglobin genes are expressed at different stages of development. The graph below shows the level of expression of various types in each stage.

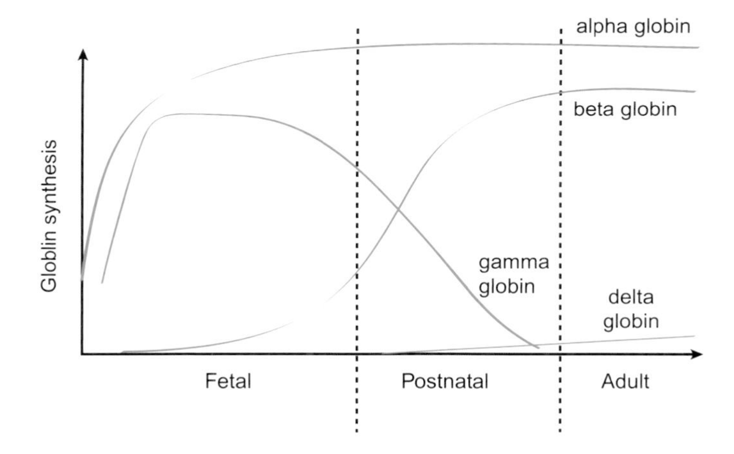

Haemoglobin A, the major type of adult haemoglobin, is a multimeric protein containing **two alpha-** and **two beta-subunits**. Each subunit is associated with a **haem group** (located within a hydrophobic pocket of the haemoglobin peptide). Hence one haemoglobin A tetramer has four haem groups. Each of the haem groups is able to associate with one molecule of molecular oxygen. A molecule of oxygen binds to the Fe^{2+} of each haem group. The binding does not change the valency of the iron atom and it remains in the Fe^{2+} state (required for the binding of oxygen). In the deoxy form, water is excluded from the hydrophobic pocket. Water would change the valency of the iron ion, and as such would remove the ability of the haem group to associate with oxygen. Binding of one oxygen molecule to the deoxyhaemoglobin tetramer causes relative movement of the subunits in the alpha1beta1 plane and enhances the binding of subsequent oxygen molecules.

Answers

34. T T T F F
35. T T F F T
36. T T T T F

37. Oxygen saturation curve of haemoglobin and myoglobin

a. Draw the oxygen saturation curve of haemoglobin and myoglobin, indicating the units of the *x* and *y* axis
b. Briefly describe how the properties of haemoglobin and myoglobin ensure that skeletal muscle is oxygenated adequately during rest and exercise

38. Regarding the binding of oxygen to haemoglobin

a. The oxygen molecule binds to the haem group
b. Oxygen displaces water from the binding site
c. Binding of oxygen converts Fe^{2+} to Fe^{3+}
d. 2,3-Diphosphoglycerate(2,3-DPG) enhances oxygen binding
e. Binding of the first O_2 enhances the binding of the second O_2

39. The following will shift the oxygen–haemoglobin dissociation curve to the right

a. Increasing pH
b. Increased temperature
c. Carbon monoxide
d. Decrease in 2,3-DPG concentration
e. Increase in carbon dioxide concentration

2,3-DPG, 2,3-diphosphoglycerateglycerate

EXPLANATION: HAEMOGLOBIN (ii)

Haemoglobin binds loosely and **reversibly with oxygen**. This is the principle by which haemoglobin is able to transport oxygen from areas of high concentration (lungs) and deliver it to areas of lower concentration (tissue). Haemoglobin becomes completely saturated only at relatively high partial pressures of oxygen (e.g. in the alveolar capillaries where PaO_2 is near 14 kPa). As the partial pressure of oxygen declines (e.g. in muscle tissues), oxyhaemoglobin readily gives up its oxygen (see diagram below). The myoglobin of resting muscle cells is able to bind oxygen at low partial pressure. Hence there is a transfer of oxygen from oxyhaemoglobin to myoglobin. Myoglobin is able to remain bound to oxygen until the partial pressure becomes very low (see diagram below). This allows myoglobin to become a store of oxygen in muscle cells. During exercise, the amount of oxygen delivered by the blood to muscle can be insufficient and the partial pressure of oxygen can become very low. When this occurs, oxymyoglobin disassociates, therefore ensuring that muscle tissue is adequately oxygenated. In continued exercise, the oxymyoglobin of muscle tissues can become completely desaturated and metabolism becomes anaerobic (38).

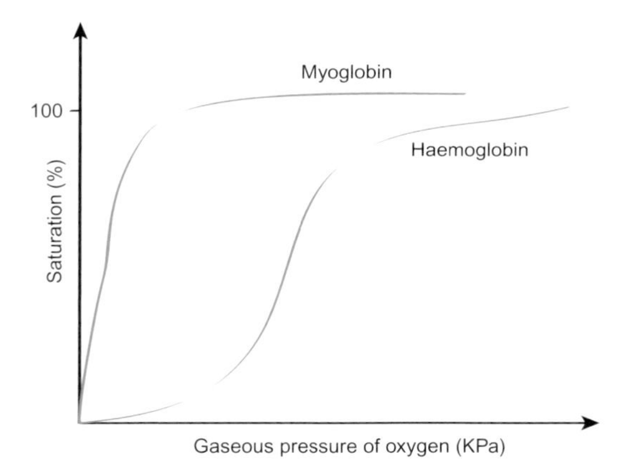

The following factors shift the haemoglobin dissociation curve to the right:

- Decreasing pH
- Increased temperature
- Increase in 2,3-DPG concentration
- Increase in carbon dioxide concentration

Carbon monoxide binds tightly to haemoglobin and has a **higher affinity** than oxygen (some 250 times more). Therefore, even when carbon monoxide concentrations are much lower than oxygen concentrations, carbon monoxide will preferentially bind to haemoglobin. Consequently, carbon monoxide can be lethal even at low concentrations. Both myoglobin (found in muscle) and fetal haemoglobin have greater affinities for oxygen than haemoglobin A.

Answers

37. See explanation
38. T F F F T
39. F T F F T

40. Theme – adhesion and extracellular matrix molecules. Pick one molecule from the list below (A–L) that fits each description (1–5)

Options

- **A.** Alpha tubulin
- **B.** Chondroitin sulphate
- **C.** Collagen
- **D.** Elastin
- **E.** Fibronectin
- **F.** Heparin
- **G.** Hyaluronic acid
- **H.** Integrin
- **I.** Laminin
- **J.** Myelin
- **K.** N-cadherin
- **L.** N-CAM

1. A transmembrane protein found in neural tissue that requires Ca^{2+} ions for homophilic binding
2. A transmembrane protein with immunoglobulin-like domains in the extracellular portion of the molecule
3. A heterodimeric class of transmembrane proteins which link the cell surface to extracellular connective tissue proteins
4. An extracellular matrix molecule with a coiled-coil structure that is able to bind to collagen and heparin sulphate
5. A protein-linked gel-like substance found in skin

41. True or false? Elastin

- **a.** Has a highly ordered tertiary structure
- **b.** Polpypeptides are covalently bonded together via lysine residues
- **c.** Is a highly hydrophilic protein
- **d.** Is rich in proline and glycine residues
- **e.** Is found in arteries and lungs

CAMs, cell adhesion molecules

EXPLANATION: EXTRACELLULAR MATRIX MOLECULES

Cadherins are **transmembrane** proteins that are linked to the cytoskeleton network inside the cell. The extracellular domain of cadherins of one cell can interact with cadherins on another cell (homophilic binding) in the presence of **Ca^{2+} ions**. They form one of the strongest intercellular links. There are many types of cadherins (e.g. N-cadherin of neural tissue).

Cell adhesion molecules (CAMs) are also transmembrane proteins attached to the cytoskeleton. The extracellular domains of CAMs contain immunoglobulin-like domains that are able to bind homophilically with other CAMs. Several types of CAMs exist that are specific for certain tissues, e.g. **E-CAM** (epithelia), **N-CAM** (neural tissue). They form weaker bonds than cadherins and do not require cations for interaction.

Integrins are heterodimeric transmembrane proteins that also interact with the cytoskeleton. The extracellular domains are capable of interacting with a wide variety of extracellular components, including laminin, fibronectin and collagen.

Laminin is a heterodimeric protein of the extracellular matrix that has a coiled-coil structure and that is able to interact with many molecules such as collagen and heparin sulphate and to bind to specific laminin receptors on cell surfaces.

Chondroitin sulphate is a **glycosaminoglycan**. These are large, highly charged molecules composed essentially of amino sugars, often linked to proteins found in the **extracellular matrix** (e.g. chondroitin sulphate, heparin). They are able to absorb huge quantities of water and thus become gel-like. Different types of glycosaminoglycans are located in different sites of the body, e.g. chondroitin sulphate is found in the skin; heparin is found in association with mast cells.

Elastin polypeptides are rich in proline and glycine residues, and are highly hydrophobic molecules. The protein adopts an unstructured, random coil configuration and individual molecules are covalently bound to each other via lysine residues. The elasticity of the molecule resides in the ability of the protein to uncoil and stretch reversibly when a force is applied. This property is particularly useful in organs that are constantly exposed to deforming forces (e.g. arteries, lungs, skin).

Answers

40. 1 – K, 2 – L, 3 – H, 4 – I, 5 – B
41. F T F T T

42. The pro alpha-chain polypeptide of collagen

a. Contains a proline at every third amino acid residue in the central region
b. Is rich in glycine
c. Adopts a left-handed helical structure
d. Is hydroxylated at proline residues
e. Is the most abundant protein in mammals

43. Fibrillar collagen (type I)

a. Has a rope-like structure
b. Is composed of two collagen alpha-polypeptide chains
c. Is held together by hydrogen bonds
d. Has a left-handed helical structure
e. Can assemble into larger structures

44. Regarding collagen

Choose one answer from the list (A–I) below that matches the description in each question. In the options below, X and Y are any amino acids

Options

A. Collagen type I
B. Collagen type II
C. Collagen type IV
D. Collagen type VIII
E. Collagen type XII
F. Glycine–X–Y
G. Glycine–X–serine
H. Proline–X–Y
I. Proline–X–lysine

1. All fibrillar collagens have this repeating amino acid sequence
2. The main type of collagen found in bone
3. The main type of collagen found in cartilage
4. A type of collagen found in the basement membrane
5. A type of collagen found mainly in vascular and corneal tissue

45. The following diseases are related to collagen abnormalities

a. Scurvy
b. Ehlers–Danlos syndrome
c. Chondrodysplasia
d. Marfan's syndrome
e. Osteogenesis imperfecta

EXPLANATION: COLLAGENS

Collagens are the most **abundant protein** found in multicellular organisms (25 per cent of total protein) and are a **major component** of skin, bone and other connective tissue. They form a family of proteins comprising at least 19 different varieties. All collagens have a repeating amino acid sequence, glycine–X–Y, where X and Y can be any amino acid residues, but are often proline or hydroxyproline. Proline, because of its ring structure, allows the molecule to fold into a **left-handed helical structure**. Proline and lysine residues are hydroxylated, which is important in the assembly of the collagen molecule Collagens can be classified as fibrillar or non-fibrillar. **Fibrillar collagens** include **type I** (found mainly in bones) and **type II** (**cartilage** and vitreous humour). The non-fibrillar collagens include the **types IV** and **VII** (found in the basement membranes) and the types VIII and X short-chain collagens (found in vascular and corneal tissue).

Fibrillar collagen is composed of **three alpha-chains** of the collagen polypeptide, has a rope-like structure and adopts a right-handed helical conformation. The structure is held together by hydrogen bonds. Hydroxyl groups of the hydroxylated proline and lysine residues form hydrogen bonds that are vital in maintaining the triple-stranded helical organization. These collagen molecules often aggregate into larger molecules called collagen fibrils, which themselves can combine to form even larger structures called collagen fibres. Many diseases result from a defect in collagen synthesis:

- **Scurvy** – Vitamin C is required as a **cofactor** for the hydroxylation of proline and glycine residues. Without hydroxylation the polypeptides are unable to form the triple-stranded helical structure of procollagen.
- **Ehlers–Danlos** syndrome – Mutations in **type III collagen** lead to the formation of fragile blood vessels and skin. The joints are also hypermobile.
- **Chondrodysplasia** – A defect in **type II collagen** results in abnormal cartilage and leads to bone and joint deformities.
- **Osteogenesis imperfecta** – This disease arises as a result of mutations in **type I collagen** that results in weak bones that fracture easily.
- **Marfan's syndrome** – This arises as a result of defects in the **fibrillin** gene that is essential for the integrity of elastic fibres.

Answers

42. F T T T T
43. T F T F T
44. 1 – F, 2 – A, 3 – B, 4 – C, 5 – D
45. T T T F T

46. True or false? Microtubules

a. Are a component of the cytoskeleton
b. Are composed of actin molecules
c. Are involved in the segregation of metaphase chromosomes
d. Are present in flagella
e. Are smaller than intermediate filaments

47. Regarding actin

a. It is present in all cells
b. The filaments of F-actin have polarity
c. The polymerization of the subunits of actin does not require ATP hydrolysis
d. Thymosin facilitates the polymerization of actin subunits
e. The polymer is a helical structure that contains 13 subunits per helical turn

48. Antibodies

a. Are produced by macrophages
b. Are composed of two heavy and two light chains
c. Hydrogen bonds keep the chains together
d. Can be cleaved by papain
e. The antigen binding domain is located on the Fc protion

49. The antigen binding site of an antibody

a. Is derived from both the heavy and light chains
b. Contain hypervariable regions
c. Interact with antigen solely by hydrogen bonds
d. Is usually heavily glycosylated
e. Can usually accommodate a peptide of 100 amino acids

50. The following associations are correct

a. IgM/primary antibody response
b. IgE/hypersensitivity response
c. IgM/monomer
d. IgA/dimer
e. IgG/pentamer

EXPLANATION: CYTOSKELETAL PROTEINS

Microtubules are made of long polymers of **alpha-** and **beta-tubulin** proteins arranged in a 13-fold ring structure and are important components of the cytoskeleton. They give mechanical support and anchor organelles to particular regions within a cell. They are essentially hollow tubes with a diameter of 25 nm (larger than intermediate filaments which are 10 nm) and are also found in cilia and flagella where they adopt the classic 9 + 2 configuration. In cilia and flagella, the microtubules are associated with dynein, which allows the microtubules to slide relative to each other, thus enabling motility.

Actin is a ubiquitous protein that accounts for some 5 per cent of cellular protein. Globular actin subunits (**G-actin**) are polymerized into filamentous actin (F-actin) which has a polarity. Although ATP is hydrolysed during polymerization, it is not necessary. Hydrolysis of ATP however greatly increases the rate of polymerization. The polymer is a helical structure with a 7 nm diameter. It contains 13 G-actin subunits per one complete helical turn. Thymosin binds G-actin and prevents polymerization.

EXPLANATION: ANTIBODIES

Antibodies (secreted by B cells) are composed of a **pair of heavy** and **light chains** that are held together by interpeptide **disulphide bridges**. Papain and pepsin are able to cleave the antibody. Digestion of the immunoglobulin with papain gives rise to two large fragments – the constant Fc fragment and the antigen-binding Fab fragment.

The antigen-binding site is derived from the variable regions of both heavy and light chains. Most of the **variability** in **amino acid sequence** in this area of each chain is concentrated around three domains called **hypervariable regions**. It is this variability that gives rise to the huge number of antibodies *in vivo*. The antigen-binding site can accommodate small peptide sequences in the region of 10–15 amino acids in length. The antigen is retained in the antigen-binding site by many forces, including hydrogen bonds, van der Waals forces and hydrophobic interactions.

The initial antibody response to an antigen is always an **IgM response**. Subsequent challenges are met with an **IgG response**. The IgM molecule is a pentamer of immunoglobulin molecules that are arranged like the spokes of a wheel, with the antigen-binding regions around the edges. IgG, in contrast, is composed of a monomeric immunoglobulin. IgA can either be a dimer or a monomer and is found in secretions (e.g. in saliva, sweat and in the gut). **IgE**, a monomer, is involved in **hypersensitivity** reactions.

Answers

46. T F T T F
47. T T T F T
48. F T F T F
49. T T F F F
50. T T F T F

51. True or false? Insulin

- **a.** Is composed of an A chain and a B chain linked by two disulphide bonds
- **b.** Is derived from two different polypeptide chains
- **c.** Is formed by the action of a protease on the prohormone
- **d.** Is synthesized as required and is not stored in the beta-cells of the pancreas
- **e.** Undergoes extensive modification on release from pancreatic beta-cells

52. Entry of proteins into the rough endoplasmic reticulum (true or false?)

- **a.** Occurs after complete synthesis of the polypeptide chain in the cytosol
- **b.** Requires specific sequences at the N-terminal
- **c.** Is dependent on the presence of polar residues at the N-terminal
- **d.** Is a receptor-mediated process
- **e.** Is an essential step in the synthesis of proteins destined for secretion from the cell

SRP, signal recognition particle

EXPLANATION: INSULIN SECRETION

Insulin is synthesized as **preproinsulin** in the **beta-cells** of the **pancreas**. The polypeptide is channelled into the lumen of the endoplasmic reticulum where extensive **modification** gives rise to mature **insulin**. The entry of insulin into the endoplasmic reticulum is coupled with synthesis on ribosomes. Proteins targeted to the endoplasmic reticulum have a signal peptide sequence at their amino (N) terminals which contains a stretch of hydrophobic residues. The synthesis of this signal peptide sequence allows the SRP to bind to both the peptide and the ribosome. Binding of the SRP halts translation and also allows the ribosome–SRP complex to associate with SRP receptors located on the cytoplasmic face of the endoplasmic reticulum. Once associated with the receptor, the SRP dissociates and translation is resumed. The growing polypeptide chain is transferred to the endoplasmic reticulum lumen via a translocation channel. Within the lumen, a signal peptidase cleaves the signal sequence from the peptide.

Once the insulin polypeptide enters the endoplasmic reticulum lumen, three intrapeptide disulphide bonds are formed. This gives rise to the proinsulin molecule. This is then cleaved at two sites by prohormone convertases to yield the A and B chain. The A and B chain remains connected by disulphide bridges. The proteolytic processing also yields a C or connecting peptide. Basic residues are removed from the B chain to yield mature insulin. Mature insulin is then packaged into secretory vesicles, where it is stored until release is demanded. Insulin does not undergo further modification once packaged into vesicles.

Answers

51. T F T F F
52. F T F T T

53. The sickle cell syndrome

a. Is an autosomal recessive condition
b. Results from a mutation in the delta-globin gene
c. Causes polymerization of haemoglobin at low oxygen tensions
d. Results in the occlusion of small blood vessels
e. Causes a polycythaemia

54. Choose one disease from the list (A–J) that arises from the defect of protein structure, synthesis or modification indicated in each question 1–5

Options

A. Creutzfeldt–Jakob disease
B. Cushing's disease
C. Cystic fibrosis
D. Diabetes mellitus
E. Ehlers–Danlos type VI disease
F. Hypercholesterolaemia
G. I cell disease (lysosomal storage disorder)
H. Sickle cell disease
I. von Willebrand's disease
J. Beta-thalassaemia

1. Imbalance in the rate of synthesis of subunits of a multimeric protein
2. Amino acid substitution in the primary sequence
3. Defect in hydroxylation of lysine residues
4. Defect in phosphorylation of mannose residues in a glycoprotein
5. Misfolding of the protein at the secondary level of organization

EXPLANATION: DISEASES ARISING FROM DEFECTS IN PROTEINS

Sickle cell syndrome results from a single base mutation in the **beta-globin** gene that results in a **glutamate** to **valine substitution** at residue 6 of the protein. Residue 6 is located at the alpha/beta-globin interface. The substitution of the glutamate with valine does not affect the structure of haemoglobin at high oxygen tensions. However at low oxygen tensions, the substitution causes a **conformational change** in **protein structure** which results in the polymerization of haemoglobin within erythrocytes. This causes sickling of the erythrocytes and these deformed cells become trapped in capillaries, causing occlusion and ischaemia. The deformed cells are removed by the spleen and consequently anaemia results (polycythaemia = increase in erythrocyte concentration).

In **beta-thalassaemia** there is reduced or no production of **beta-globin** protein due to a gene lesion. Severe anaemia results.

Ehlers–Danlos type VI disease is caused by a deficiency in lysyl hydroxylase that results in defective **hydroxylation** of **collagen**. Hydroxylation of lysine residues of collagen is required for the intermolecular crosslinks between collagen molecules that stabilize the collagen fibrils.

Acid hydrolases are glycoproteins that are targeted to the lysosome by phosphorylation of mannose residues in the Golgi apparatus. In I cell disease there is a defect in the modification of the mannose residues and hence acid hydrolases are lacking in the lysosomes and instead are secreted from the cell.

Creutzfeldt–Jakob disease is a **prion protein disease**. PrPc is a normal cellular protein with alpha helices and no beta sheets and is sensitive to proteases. It is believed that **abnormally folded protein PrPSc** acquires beta sheets and becomes **protease resistant**. It accumulates in lysosomes and eventually causes the **rupture** of **lysosomes** and the **cell itself**. The mutant protein is also able to induce abnormal folding of normal PrPc proteins and thus propagation is achieved.

Answers

53. T F T T F
54. 1 – J, 2 – H, 3 – E, 4 – G, 5 – A

55. Protein analysis. The following questions relate to the techniques available for the analysis of proteins. Choose one appropriate analytical technique from the list (A–I) below for each of the descriptions in the questions

Options

A. Affinity chromatography
B. Gel filtration chromatography
C. Immunofluorescence assay
D. Ion-exchange chromatography
E. Isoelectric focusing
F. Protein sequencing
G. Polyacrylamide gel electrophoresis in the presence of an anionic detergent like sodium dodecyl sulphate (SDS-PAGE)
H. Polymerase chain reaction
I. Western blotting

1. The separation of proteins purely by mass using an electric field
2. The separation of proteins by mass without using an electric field
3. The separation of proteins by their charge in the presence of an electric field
4. The separation of proteins by their charge in the absence of an electric field
5. The separation of proteins using antibodies attached to a matrix

SDS-PAGE, sodium dodecyl sulphate polyacrylamide gel electrophoresis

EXPLANATION: PROTEIN ANALYSIS

Proteins can be **analysed** by virtue of their **charge** and/or their **molecular mass**. **SDS-PAGE** is a technique that separates protein purely on size. Proteins to be analysed are loaded onto a matrix (**polyacrylamide gel**) that contains the anionic detergent SDS. SDS coats the proteins and makes them negatively charged. An electric current is passed through the gel, and proteins migrate according to their mass; low mass proteins migrate the furthest.

Proteins can also be separated by **mass** in the absence of an electric field. In **gel filtration chromatography** proteins are loaded onto a porous matrix and washed through using a solvent. While large proteins do not enter the pores and are washed quickly through, smaller proteins are able to enter the pores and take longer to traverse the matrix. The ease with which different proteins are able to enter the pores is dependent on their size, and the easier it is for a protein to enter the pore, the greater its retardation in the matrix.

The **charge** of a protein can also be used to separate proteins. At a particular pH, a protein will have a certain net charge which is positive or negative. An oppositely charged matrix can be used to bind the protein at this pH. By changing the pH, the charge of the bound protein can be changed so that it no longer binds to the matrix. At the correct pH the bound protein will elute from the column. This is the principle of **ion-exchange chromatography**.

At a **particular pH** a protein will have **no net charge**. This is known as the **isoelectric point**. In isoelectric focusing, proteins are separated (using an electric field), through a matrix in which a pH gradient has been established. Initially the proteins will migrate towards either electrode because of their charge. In doing so they will also migrate through the pH gradient and so their charge will begin to diminish. Eventually, the proteins will migrate to an area in which the pH corresponds with the protein's isoelectric point. At this point the proteins will have no net charge and hence will not move in the electric field. Hence different proteins with different isoelectric points will migrate to different points within the matrix.

Proteins can also be separated by **affinity chromatograpy**. Here a matrix is linked to a substance that specifically binds to the protein of interest, e.g. a specific antibody or a ligand for a protein or an enzyme substrate analogue. In such systems, only the protein of interest will bind and can be purified.

Answers

55. 1 – G, 2 – B, 3 – E, 4 – D, 5 - A

56. Case study

At the emergency department, a 65-year-old man is found on examination to be unwell with generalized oedema (legs, sacral, groin and face) and ascites (fluid in the peritoneum). He has a long-standing history of alcohol abuse. The results of some of his blood tests are shown below. The values in brackets are normal values.

Electrolytes		
Na^+	142 mmol/L	(135–145)
K^+	4.1 mmol/L	(3.5–5)
Liver function tests (LFT)		
Albumin	18 g/L	(35–50)
Alkaline phosphatase	587 U/L	(30–300)
Alanine aminotransferase	168 U/L	(5–35)
Bilirubin	154 μmol/L	(3–17)

a. How is the distribution of extracellular water between the vascular and extravascular (interstitial) compartments controlled?
b. What might have caused the oedema and ascites in this patient?

EXPLANATION: OEDEMA AND ASCITES

Water is able to **move freely** between **blood vessels** and the **interstitium**. The distribution of extracellular water between the two spaces is determined by the **equilibrium** between the **oncotic pressure** and the **hydrostatic pressure** in the two spaces, as shown in the diagram below. The net hydrostatic pressure results in water leaving the capillary lumen. **Plasma** has a **higher oncotic pressure**, due to its high protein (mainly albumin) and hence the **net oncotic pressure** drives **water** into the **capillaries**. However, the **net hydrostatic pressure** is higher than the net oncotic pressure and hence there is a **gradual loss** of water from the blood vessels. This excess fluid is normally absorbed by the lymphatic system and is returned to the blood (56a).

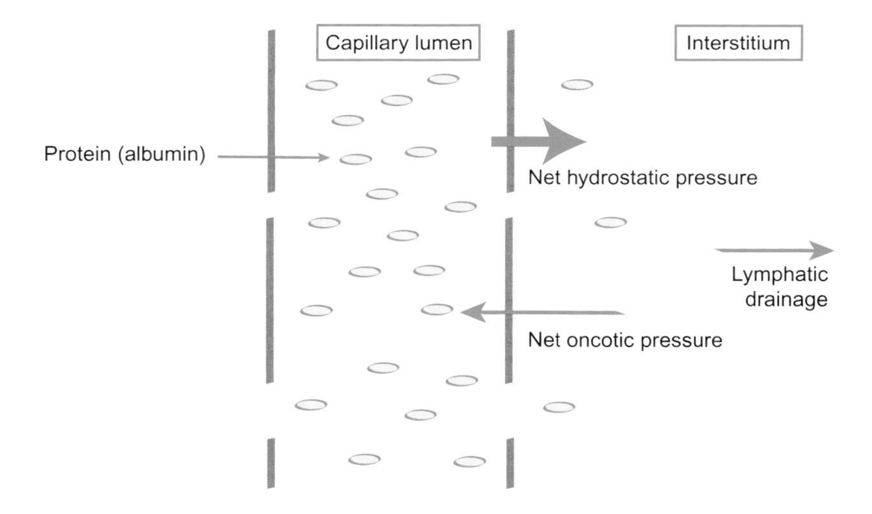

Some 10–15 g of albumin is produced by the adult liver every day and at times of need, this can be doubled. The liver is unable to store albumin. The normal plasma concentration of albumin is between 35 and 50 g/L (~0.6 mmol/L), and it has a long half-life in the plasma ($t_{1/2}$ of 20 days). **Eighty per cent of the effective osmotic pressure of plasma can be attributed to albumin and hence it is vital in controlling water distribution between the compartments**. In this patient the history of alcohol abuse and the elevated liver enzymes and bilirubin indicate that he has advanced liver disease. The ability of his liver to produce albumin has diminished and consequently the oncotic pressure of his plasma has fallen. Hence not enough fluid is reabsorbed back into the bloodstream and oedema develops. Cirrhosis of the liver also leads to portal hypertension, causing an increase in the hydrostatic pressure within these blood vessels. This increase in hydrostatic pressure, coupled with a reduced plasma oncotic pressure, leads to the development of ascites (56).

Answers

56. See explanation

INDEX

9 + 2 arrangement of microtubules 45, 55
A alpha fibres 35
abdominal surgery case study 24–5
acetylcholine 14–15, 25
acetylcholinesterase 19, 25, 27
acid hydrolases 141
actin 52–5, 66-7, 136-7
actinin 54–5
action potentials 30–5
active transport 10–11
adenine 73, 74, 75
adenosine diphosphate (ADP) 67
adenosine triphosphate (ATP) 8–11, 67, 72, 75
adenylate cyclase 23
adherens junctions 57
ADP *see* adenosine diphosphate
adrenoleukodystrophy 47
affinity chromatography 142–3
agarose gel electrophoresis 92–3
agonists 22–3
albumin 127–8, 144–5
alpha2 adrenoceptors 16–17
alpha globin 128–9
alpha helix 118–19
alpha helix bundles 120–1
alpha tubulin 137
amino acids 83, 110–15, 123, 126–7
amniocentesis 92–3
amphipathic molecules 2–5
amplification of DNA 91, 94–5
anaphase 59
androgen receptors 16–17
antagonists 22–3
antibiotics 82–3
antibodies 136–7
anticodons 80–1
antigen binding sites 136–7
antiporters 8–9
apoprotein B100 14–15
apoptosis 60–1
arachidonic acid 6–7
aromatic side chains 111–12
ascites 144–5
asymmetry of organs 45
ATP *see* adenosine triphosphate
atracurium 18–19, 25
atropine 18–19
autoimmune diseases 16–17
autosomal dominant genes 15–17, 100–1
autosomal recessive genes 104–5
axonemal dynein 45
axoplasmic flow 55
AZT 27

bacteria, lack of organelles 39
basal lamina 56–7
base pairing 72–5
basophils 68–9
benzene rings 113
beta adrenergic receptors 16–17, 22–3
beta–alpha–beta folds 120–1
beta barrels 120–1
betablockers 19
beta globin 128–9, 140–1
beta sheets 118–9
beta thalassaemia 140–1
beta tubulin 137
bilayers, phospholipid 2–5
blood 68–9, 126–131, 144–5

cadherins 132–3
calcium 20–1, 30–1, 55
calcium channels 35
carbon monoxide poisoning 131
carboxyl groups 110–11
cardiac muscle 34–5
catalase 48–9
cell adhesion molecules 132–3
cell membranes 2–35
cells
  see also individual cell types
  cycle 58–61
  division 62–3
  intracellular structures 48–51
  junctions 56–7
  nuclei 38–9
cell surface receptors 16–17
cellular signalling 20–1
centriole 50–1
centrosomes 48–9, 52–3
ceramide 20–1, 22–3
chloride ion concentrations 30–1

cholesterol 2–3, 5, 14–15
chondrodysplasia 134–5
chondroitin sulphate 132–3
chorionic villus sampling 92–3
chromatids 59
chromatography 142–3
chromosomes 58–9, 84–5, 100–1, 102–3
cilia 44–5, 54–5
ciliated simple columnar epithelium 64–5
cirrhosis of the liver 144–5
cisternae 40–1, 48–9
clastogens 98–9
clathrin 51
cloning DNA 90–1
codons 80–3
coiled coil structures 119, 120
collagens 134–5
competitive antagonists 22–3
competitive reversible inhibitors 26–9
concentration gradients 8–11
conduction, saltatory 31, 35
contraction of muscles 52–3, 55, 66–7
covalent bonds, proteins 116–7
COX pathways 6–7
Creutzfeldt–Jakob disease 140–1
cyclic AMP 20–1, 22–3
cyclic GMP 20–1
cyclic nucleotides 72, 75
cyclin dependent kinases 60–1
cyclo oxygenase pathway 6–7
cystic fibrosis 45, 94–5, 104–5, 106–7
cytochrome P450 monooxygenase 7
cytokinesis 59
cytosine 73–4
cytoskeleton 52–5, 136–7
cytosol 50

DAG *see* diacylglycerol
delta globin 128–9
deoxyribose sugar 73
depolarization 32–5
desmosomes 56–7
diacylglycerol (DAG) 20–1, 22–3
diffusion 8–11
diploid cells 62–3, 84–5
disulphide bonds 42–3, 123, 125, 137
dizygotic twins 100–1, 102–3
DNA
  amplification 91, 94–5
  analysis 92–3
  bacteria 39
  cloning 90–1
  damage 137–41
  mitochondria 39, 44–5
  mutations 92–3, 95, 96–9, 102–5
  nucleotides 72–5
  packaging 88–9
  polymerase chain reaction 94–5
  polymerases 88–9
  replication 39, 58–63, 88–9
  transcription 76–7
dominant genes 100–1
Down's syndrome 68–9, 85, 102–3
dynamin 50–1

editing mRNA 79
edrophonium 25
Edward's syndrome 102–3
Ehlers–Danlos syndromes 47, 134–5, 140–1
eicosanoids 6–7
elastin 132–3, 135
electrostatic interactions 117
endocytosis 50–1
endoplasmic reticulum 42–3, 81, 138–9
enzymes 26–9
eosinophils 68–9
epidermis 62–3
epithelia 44–5, 64–5
epithelial basement membrane 56–7
erythrocytes 39, 68–9
euchromatin 86–7
eukaryotes 38–9, 76–81, 84–5, 89
exocytosis 50–1
exons 78–9
extracellular matrix molecules 132–3
extracellular water 144–5

facilitated diffusion 9–11
familial hypercholesterolaemia 14–15
family trees 102–5
fatty acids 4–5
fetal genetic analysis 92–5

fetal haemoglobin 128–9, 131
fibrillar collagen 134–5
fibrinogen 126–7
flagella 52–3, 54–5
flippases 3
fluid mosaic model 2–5

gamma globin 128–9
gap junctions 56–7
Gaucher's disease 46–7
gel electrophoresis 92–3, 142–3
gel filtration chromatography 142–3
genetics 86–7, 92–3, 100–1, 102–3
genomes 84–5
glycolysis 49, 75
glycopyrrolate 24–5
glycosaminoglycans 133
glycosylation 42–3, 122–3
Golgi apparatus 40–1, 48–9, 50–1
G phases 58–61
G protein coupled receptors 19–21, 22–3
gradients, membrane transport 8–11
granulocytes 69
Graves' disease 16–17
guanine 73, 74
guanosine triphosphate (GTP) 11, 20–1, 23, 72, 75

H^+ ATPase pump 48–9
haemoglobin 128–31
haploid cells 58–9, 62–3, 84–5
hepatocytes 43, 48–9
hereditary disorders 92–3, 102–7
heterochromatin 87
heteronuclear RNA 78–9
histamine 69
histology 62–9
histones 87, 89
HIV 27
hormones 86–7
human genome 84–5
hydrogen bonds 73, 116–17, 118–19
hydrostatic pressure 145
hydroxylation of proteins 123, 125

I cell disease 140–1
immunoglobulins 126–7, 136–7
inheritance 92–3, 100–7
inhibition, enzyme 26–7
insertional mutations 96–7, 99
insulin 138–9
insulin receptor 14–15, 16–17, 18–19
integrins 132–3
intercellular junctions 56–7
intermediate filaments 52–3
interphase 58–9
intracellular structures 48–51
introns 78–9
ion exchange chromatography 142–3
ionic bonds, proteins 116–17
irreversible antagonists 22–3
irreversible inhibitors 26–7
isoprenaline 18–19

junctions 56–7

K^+ *see* potassium
Kartagener's syndrome 44–5, 55, 68–9
kinetics, enzyme 28–9
kinetochores 59
Kleinfelter's syndrome 102–3

laminin 132–3
LDL *see* low density lipoprotein
leprechaunism 16–17
ligands, receptors 14–15
Lineweaver–Burk plot 28–9
lipid anchored proteins 12–13
lipids 2–25, 48–9
lipofuscin 49
lipoxygenase pathways 7
liver disease 144–5
low density lipoprotein (LDL) receptor 14–15
lymphocytes 69
lysophospholipids 3
lysosomal storage disorders 46–7, 68–9
lysosomes 46–9, 50

Marfan's syndrome 134–5
mast cells 69
meiosis 58–9, 62–3
membranes 2–35
messenger RNA (mRNA) 78–83, 87

metaphase 59
metoprolol 18–19
Michaelis–Menten kinetics 10–11, 29
microtubules 44–5, 49, 52–3, 55, 136–7
microvilli 50
mitochondria 44–5, 49–51
mitogens 98–9
mitosis 58–9, 62–3
mivacurium 25
monocytes 69
monotopic integral membrane proteins 12–13
monozygotic twins 100–1, 102–3
M phase 60–1
mRNA *see* messenger RNA
mucus, cilia 45
multiple sclerosis 68–9
muscarinic acetylcholine receptor 16–17
muscles 24–5, 35, 52–3, 55, 66–7
mutagens 98–9
mutations 92–3, 95, 96–9, 102–5
myasthenia gravis 16–17
myelination 31, 34–5, 62–3
myofibrils 54–5, 67
myoglobin 130–1
myosin 52–3, 54–5, 66–7

Na^+ *see* sodium
Na^+/K^+ATPase pump 8–9, 11, 18–19
neostigmine 18–19, 25, 27
nerves 30–5, 62–3
neuromuscular blocking agents 24–5
neurotransmitters 113
neutrophils 68–9
nicotinic acetylcholine receptor 14–15, 16–17, 25
nodes of Ranvier 31,35
noncompetitive enzyme inhibitors 28–9
NSAIDS 7
nuclei 38–9, 50–1
nucleic acids *see* DNA; RNA
nucleolus 50
nucleosomes 88–9
nucleotides 72–5, 96–7

odema 144–5
Okazaki fragments 88–9
oligodendrocytes 62–3
oncogens 98–9
oncotic pressure 145
organelles 38–9, 48–51
  see also individual organelles
osteogenesis imperfecta 134–5
oxidative phosphorylation 45, 48–9
oxygen binding of haemoglobin 130–1

pancuronium 23, 25
partial agonists 23
passive transport 8–11
Patau's syndrome 102–3
pathology 68–9
PCR *see* polymerase chain reaction
pedigree charts 102–3, 104–5
pemphigus 47, 68–9
penetrance of genes 100–1
peptides and peptide bonds 116–17, 124–5
peripheral membrane proteins 12–13
peritoneum 144
peroxisomes 48–9, 50
pH 30–1, 44–5, 48–9
phagocytosis 50–1
phosphatidic acid 22–3
phosphatidylcholine 4–5
phospholipase A_2 7
phospholipase C 20–1
phospholipids 2–5
phosphorylation 19, 45, 48–9, 122, 125
pinocytosis 51
plasma 126–7
plasma cells 68–9
plasma membranes 12–13
plasmid vectors 90–1
point mutations 96–7, 99, 102–5
polyacrylamide gel electrophoresis 142–3
polymerase chain reaction (PCR) 93, 94–5
polymerases 76–7, 88–9
polypeptides 80–3, 118–19
polyploidy 84–5
polytopic integral membrane proteins 12–13
posttranslational protein modification 122–3, 138–9
potassium channels 32–3, 35
potassium ion concentrations 30–1
prenatal genetic testing 92–5, 107

prepro insulin 138–9
primary active transport mechanisms 11
primary structure of proteins 118–9
prion protein diseases 141
prokaryotes 38–9, 76–7, 80–3
prometaphase 58–9
prophase 58–9
prostacyclin 6–7
prostaglandins 7
prostanoids 6–7
protein kinase C 20–1
proteins
  amino acid side chains 110–15
  analysis 142–3
  in blood 126–31, 144–5
  cytoskeletal 52–5, 136–7
  defect diseases 140–1
  entry to endoplasmic reticulum 138–9
  gene expression 86–7
  membrane 12–13
  posttranslational modification 122–3, 138–9
  sorting 43
  structure 116–21
  synthesis 42–3
  translation 80–3
  types 132–9
pseudostratified epithelia 64–5
purines 72–3
pyridostigmine 25
pyrimidines 72–3

receptors 14–25
recessive genes 100–1, 102–3, 104–5
recombinant DNA 90–1
relaxants, muscle 24–5
replication 88–9
resting membrane potential 31, 34–5
restriction enzymes 90–1
retinitis pigmentosa 16–17
reversible competitive antagonists 22–3
reversible competitive inhibitors 26–9
ribose sugar 73
ribosomes 43, 78, 79, 80–3
RNA 73, 76–83, 87
rocuronium 25
rough endoplasmic reticulum 50–1

saltatory conduction 31, 35
sarcomeres 55, 66–7
Schwann cells 62–3
scurvy 134–5
SDS PAGE analysis 142–3
secondary active transport mechanisms 11
secondary structure of proteins 118–19
second messengers 20–1, 22–3
sex chromosomes 84–5, 100–1, 102–3
sickle cell syndrome 102–5, 140–1
side chains of amino acids 110–16
signalling, cellular 20–1
simple columnar epithelium 64–5
simple cuboidal epithelium 64–5
simple diffusion 8–11
simple epithelia 64–5
simple squamous epithelium 64–5
situs inversus 45
skeletal muscle 66–7
sodium channels 32–3, 35
sodium ion concentrations 30–1
Southern blotting 92–3
S phases 58–63
steroid hormones 86–7
stratified epithelia 64–5
striated muscle 34–5
super secondary structure of proteins 118, 120–1
surgery, muscle relaxants 24–5
symporters 8–9

Tay–Sachs disease 46–7, 49, 68–9
telophase 58–9
teratogens 98–9
tertiary structure of proteins 118–19
thromboxanes 6–7
thymine 73, 74
thyroid stimulating hormone (TSH) receptor 17
tight junctions 56–7
transcription 76–7, 86–7
transfer RNA (tRNA) 80–1
transitional epithelium 65
translation 80–3, 86–7
translocation 42–3
transmembrane proteins 132–3
transport, membrane 8–11
triplet repeat expansion 102–3

trisomy 84–5, 102–3
tRNA *see* transfer RNA
tropomyosin 54–5
troponin 67
TSH *see* thyroid stimulating hormone
tubulins 137
Turner's syndrome 102–3
twins 100–1, 102–3
tyrosine kinase receptors 14–15, 18–19

uniporters 8–9
uracil 73, 74, 75
UTP 72, 75

Van der Waals attraction 116–17
vecuronium 25
vesicles 50–1

white blood cells 68–9

X-linked recessive traits 100–1, 102–3, 104–5

zwitterions 110–11